Writing in the Technical Fields

IEEE Press
445 Hoes Lane, PO Box 1331
Piscataway, NJ 08855-1331

Writing in the Technical Fields

A Step-by-Step Guide for Engineers, Scientists, and Technicians

Mike Markel
Boise State University

Sponsored by the IEEE Professional Communication Society

The Institute of Electrical and Electronics Engineers, Inc., New York

This book may be purchased at a discount from the publisher when ordered in bulk quantities. For more information contact:

IEEE PRESS Marketing
Attn: Special Sales
P.O. Box 1331
445 Hoes Lane
Piscataway, NJ 08855-1331
Fax 908-981-8062

Printed in the United States of America

10 9 8 7 6 5 4 3 2 1

ISBN 0-7803-1036-5 (pbk) **IEEE Order Number: PP03855**
ISBN 0-7803-1059-4 (case) **IEEE Order Number: PC03855**

Library of Congress Cataloging-in-Publication Data

Markel, Mike
Writing in the technical fields: a step-by-step guide for engineers, scientists, and technicians / by Mike Markel.
p. cm.
Includes bibliographical references and index.
ISBN 0-7803-1036-5
1. Technical writing. I. Title.
T 11. M3465 1994 93-26817
808′. 0666—dc20 CIP

To R. J. and David

Contents

Preface

If you are an engineer, scientist, or technician, you already know that writing is critical to your success. If you can write a memo, a letter, or a proposal that makes your point and motivates your audience, you're a valuable asset to your company. If you can't, you're much less valuable.

That's the way it is now, and the ability to write well is going to become even more important each year as communication technology improves: you will write more often to more people. This book explains a commonsense approach to workplace writing that can help you write faster and better. With a little practice, you will find that the *need* to write is really an *opportunity* to showcase your technical skills

The key to effective writing is learning a system that takes advantage of the way the mind works. Without a system, every writing task becomes an exercise in frustration. You stare at the blank screen for what seems like hours, and when you finally *do* write something, it isn't what you want to say. This book describes a simple but effective system for any kind of writing on the job.

Organization of the Book

The book is divided into two parts: techniques and applications.

Part I, on the techniques of workplace writing, is a summary of the best current thinking on how to arrange words and graphics on a page to make it easy for your readers to understand what you have to say. Chapter 1 discusses some of the major reasons people have trouble writing and then suggests ways to become more comfortable as a writer. The most important step is to realize that writing is hard not only for you but for everyone, that it takes much more time than you want to devote to it, and that the results are never perfect. With these assumptions understood, you can stop feeling frustrated and disappointed in yourself and instead put your energies into creating a document that gets the job done.

Chapters 2 and 3 discuss the writing process. First you analyze your audience and define your purpose; then you put together a plan for the document. After you make sure your boss agrees with your plan, you generate information, organize it, write the draft quickly, and spend as much time as you can revising it. You set it aside for as long as possible, pick it up, and revise it even more. This process—or some variation on it—works regardless of the kind of document you are writing.

Chapter 4 deals with word processing—how to use the computer during the different stages of the writing process. It also addresses spell checkers, thesaurus programs, and style programs. The computer is a great tool, as long as you realize it is only a tool, not the brain behind the writing.

Chapters 5 through 8 concentrate on improving the coherence of your writing—the way it hangs together—and on improving paragraphing, sentence construction, and word choice. Chapter 5 covers formal elements such as titles, headings, lists, introductions, and conclusions. Chapter 6 focuses on techniques of writing better paragraphs: providing an overview in the topic sentence, organizing the body clearly, and using transitional devices. Chapter 7 covers sentence construction, including suggestions on how to make them concise, clear, and powerful. Chapter 8 concentrates on individual words and phrases, focusing on choosing the simple, clear word and avoiding unnecessary jargon, euphemisms, clichés, and sexist language.

Chapters 9 and 10 discuss the visual elements of writing. Chapter 9 explains some basic principles of graphics: knowing when to use them, determining what kind to create, making sure they are honest and clear, determining where to put them, and linking them to the text. Chapter 10 is an overview of page design: the art of arranging words and visuals on a page so that it is attractive and easy to understand. The chapter concentrates on white space, columns, and type.

Part II of the book consists of eight chapters of advice on creating common kinds of workplace documents: letters, memos, minutes, procedures, manuals, proposals, progress reports, and completion reports. (I don't discuss electronic mail, which is a medium rather than a kind of document, or online documentation, which is such a complex topic that it requires a full-length book, such as William Horton's *Designing and Writing Online Documentation* [John Wiley, 1990].) As you read these discussions, keep in mind that workplace writing is not a science; there are no specifications or codified rules on which everyone has agreed. You won't find a document that explains, for example, how everyone should write a proposal; you won't even find complete agreement on what a proposal is. If you are lucky, your company or a professional organization in your field has published standards on what they want to see in a proposal. In most cases, however, this kind of information is not written down; you have to look around and find good examples of the kind of writing your readers are looking for and then try to figure out, from the finished product, how to put it together.

Because there are so many approaches to writing the common kinds of documents, Part II does not seek to define and exemplify all of them. This is not a book of models for you to retype, such as *100 Business Letters for Every Occasion.* Rather, Part II seeks to explain the basic strategies behind the different documents. For instance, in discussing memos, I'm less interested in the details of how the subject line is displayed in your company than in what a subject line is supposed to accomplish. In discussing completion reports, I'm less interested in what your company calls them than in the kinds of questions you want to answer in writing a report at the end of a project.

The book concludes with six appendices: a set of writing checklists; a handbook that reviews basic grammar, punctuation, and mechanics; a guide to commonly misused words and phrases; guidelines for nonnative speakers of English; guidelines for writing to nonnative speakers of English; and a bibliography.

How to Use This Book

Unless you've taken a good, comprehensive course in technical communication in the 1990s, much of the material in Part I will probably be new to you. Therefore, I recommend that you read all of Part I; it contains advice that you can put to use right away in all your writing.

Part II, on the common kinds of documents, is more of a reference. When you have to write a report and you know it is supposed to include an executive summary, Chapter 15 is the place to turn. Before you write your next letter, spend a few minutes reading Chapter 11. Part II is designed to help you understand the strategies involved in creating different kinds of documents; it will set you on the right track. But remember that workplace writing is ultimately local: the interests and needs of your readers should take precedence over any advice in this or any other book.

Appendix A is a set of checklists that serve as a review of the main points of the book. You should revise them, deleting those items that don't pertain and adding others that are missing.

Appendix B, the review of grammar, punctuation, and mechanics, is a brief look at the problems that are most common in workplace writing. Although it is true that punctuation and mechanics problems are not the main causes of ineffective workplace writing, they can undermine your credibility and professionalism. If you know that your writing suffers from these superficial problems, Appendix B will go a long way to helping you. Keep in mind, however, that a full-size handbook, such as Diana Hacker's *A Writer's Reference* (St. Martin's, 1992), is much more comprehensive.

Appendix C, the review of commonly misused words and phrases, is for everyone who cannot remember the difference between "affect" and "effect." As is the case with punctuation and mechanics, misusing words and phrases is less likely to confuse your readers than to undermine your credibility and professionalism.

Appendix D, guidelines for speakers of English as a second language, discusses some of the inscrutable aspects of the world's most difficult major language. A few pages won't make English simple, logical, and consistent, but they might help to some degree. The appendix begins with a reference to an excellent book on the subject.

Appendix E, guidelines for writing to speakers of English as a second language, offers ten tips to help you communicate with people who are not completely fluent in the language. With a little care, we can all reduce the chances that our message will be misunderstood.

Appendix F is a selected bibliography, covering such subjects as general writing, technical writing, proposals, word processing, and graphics.

No book, regardless of how good it is, can turn you into an excellent writer. You have to write and keep writing. If your company has a technical writing or technical publications department, get to know the people there. They have many resources that they would be willing to share with you. And they would also be happy to take a look at a draft, for they know that anything they can do to help you will just make their own jobs that much easier in the future.

For some twenty years I have used the system described in this book in my consulting and university teaching. This system has given thousands of professionals the skills and self-confidence to cut down the time and aggravation that go into writing and to improve dramatically the effectiveness of their finished products. I hope it will work for you, too.

Acknowledgment

Appendix B, which first appeared in my book *Technical Writing: Situations and Strategies* (3rd ed.), is printed courtesy of St. Martin's Press.

Mike Markel
Boise, Idaho

Part I

Techniques

Chapter 1

Introduction

Writing on the job is never easy, but it doesn't have to be as difficult as most people make it. Why does writing cause so much anxiety and frustration, and what you can do about it? What are the essential characteristics of effective workplace writing? These are the questions this chapter answers.

Treat Workplace Writing as a Craft, Not an Art

Over the years, dozens of people have told me, in one way or another, that they feel a little embarrassed and inadequate because they find it difficult and time consuming to write. What's more, they are often dissatisfied with the results. "I'm sure you don't have these problems," they tell me earnestly. They couldn't be more wrong.

Perhaps the biggest hurdle for most writers is a serious misunderstanding of the nature of writing. They don't realize that writing is difficult and time consuming for everyone. And they don't understand that the only writer who feels fully satisfied with the quality of the finished product is a writer with low standards.

If you were to survey even the most experienced writers, they would probably say that they don't like to write. They might add that sometimes, on a good day, they like to have written—that is, sometimes they can look at what they have made and see that they accomplished quite a bit. There are always problems that remain to be fixed, and sometimes they have to start over, but at least they have created something that wasn't there when they started. That's all they ask of themselves, and it's all you should ask of yourself.

Don't try for perfection when you start to write. The conditions under which you work pretty much rule it out. You don't have all the information you need. You don't have enough time to polish your writing. And you probably get interrupted a lot. Forget about creating the perfect document; try for "excellent" or, maybe, "very effective." Although you can master a number of techniques for improving your writing, it will never turn out flawless.

What is the best way to write on the job? There is no single method; people are just too different. Over the years, we each develop our own ways of carrying out all the tasks we do at work, and writing is no exception. Some people cannot write except on a word processor; others require legal pads and pencils. Some write only in the early morning; others can't accomplish anything until everyone else has left the office.

Despite these individual differences, however, we are beginning to learn quite a bit about successful workplace writing. This book condenses a considerable body of research—and two decades of my own experience consulting with individual writers—into a few basic techniques of workplace writing. If you don't already use some of these techniques, try them out. They may be effective, or at least you may be able to adapt some of them to your own working methods.

Unlearn What You Learned in School

One of the reasons many people have trouble writing on the job is that much of what is taught in school about solving problems and communicating results doesn't apply in the working world. Our writing teachers don't deserve all the blame, although many of them still preach the prohibitive rules—such as never end a sentence with a preposition—that don't make much sense in the working world. Rather, the problem has to do with how schools mold and form the ways we think, which in turn affects the way we approach workplace writing.

In school we learned that the projects we work on were thought up by a teacher to keep us busy, that they have little or nothing to do with real problems. It started in grade school: one train leaves Chicago at noon, traveling eastward at 95 miles per hour, and another train leaves New York at one o'clock, traveling westward on the same track at 90 miles per hour. Our job was to figure out where the two trains would collide. But if they were really hurtling toward disaster somewhere near Pittsburgh, wouldn't it be smarter to get on the radio and try to prevent the crash? Even though our teacher explained that we were doing an exercise to help us learn a set of concepts and mathematical techniques, what we heard was that problems and solutions do not always link up.

In school we learned that our writing should be "at least": at least 1,000 words, at least 10 pages (double spaced), at least something. What the teacher was doing, of course, was asking us to examine the problem in a certain amount of detail. A 2-page report was meant to be a quick overview; a 10-pager was to be more detailed and get into more complex ideas. But what we heard was that the report on *Macbeth* had to be 1,000 words long. Knowing that even if we were lucky we had only about 500 words to say about the Scottish king, we learned how to start sentences with phrases such as "It is indeed a not unsupportable contention that . . ." No wonder a lot of workplace writing sounds as if it were created by people who studied English while growing up on another planet.

In school we learned to tell our reader even the smallest details that led us to our results. The code phrase in math class was "Show your work!" We couldn't just provide the correct answer; that could be attributed to dumb luck or cheating. And even if we got the answer wrong, we might receive partial credit if we could show we used the right methods. The teacher, we knew, would read everything carefully; that's what teachers like to do.

In school we learned to pad our assignments with the biggest words and the most complicated theories we knew. The teacher knew the subject better than we did, of course, and the purpose of the assignment was to show that we knew it too. Never did we have to explain something to someone who really didn't understand what we were trying to say.

In short, we became accustomed to writing to someone who already knew what we had to say but who for some reason wanted us to write long, complicated, detailed answers that solved nonexistent problems.

Focus on What You Know About Writing at Work

If you work for the average company, you know that this approach to solving problems and communicating your findings doesn't work. Most of the people you write to don't have the knowledge, the time, or the interest to read long, complicated documents.

Because of the knowledge explosion over the past few decades, you can't assume that most of your readers know much about your subject. Technical people at your company are busy learning their own subspecialties. And most managers, even those who started out as technical people, can't even hope to stay current with all the fields their workers are researching. For most managers, it's a full-time job keeping up with the business end of the operation: hiring and performance reviews, long-range planning, quarterly status reports and sales projections, government regulations, and so forth.

Most managers couldn't find the time to read all you've written even if they could understand it; too many documents are landing on too many desks. Photocopiers have contributed quite a bit to the spread of documents in the working world, and electronic mail is speeding up the process even more, because it doesn't take any more time or cost any more money to send the document to a hundred people than to just one.

Even if managers could find the time to read the whole document, they would probably choose not to. Their interest in the document is different from yours. From your point of view, all the information is important. (After all, you did the work because it was necessary, not to keep yourself busy.) Managers realize that the full story is crucial for carrying out the recommendations and documenting the project, but they probably are not very interested in *how* you did what you did; they assume that your working methods are accurate and professional.

Rather, managers care most about your conclusions and recommendations—in other words, what you found out and what you think ought to be done about it. Unless these sections look bizarre, most managers are perfectly content to skip over the main body of the document.

Characteristics of Effective Workplace Writing

As I will discuss in the next chapter, effective workplace writing meets the needs of its particular audience. To a large extent, your success as a writer depends on how well you have analyzed what your audience needs and prefers. In general, however, it is possible to isolate a number of characteristics of effective workplace writing regardless of the audience. This section briefly describes eight characteristics.

Honesty

Above all, workplace writing should be honest. Honesty means a number of things—some obvious, some not so obvious. One obvious point is to acknowledge your use of other people's ideas or words by citing them, using whatever documentation system is appropriate in your field.

The issue of honesty in writing is often complicated by the fact that the person doing the writing does not always have the final say about what gets written. Still, writers should take responsibility for what they create. If you realize you are being asked to do something dishonest or unethical—such as lying or distorting information—you are responsible for trying to remedy the situation by appealing to the best instincts and interests of your supervisor. If that doesn't work, you should keep going up the ladder. If nobody in your organization sees things your way and you are confident that a serious problem is occurring or might occur, then it is time to consider blowing the whistle.

Your first responsibility as a writer is not to lie. For instance, if your company manufactures disk drives with an expected life of 150,000 hours, it is, of course, wrong to state that they last 200,000 hours. And you shouldn't misrepresent reality, such as by suggesting that a product design is the result of sophisticated market-research techniques, when all that really occurred is that you had a brief conversation with someone from the marketing department.

A more difficult aspect of honesty concerns what you don't say. For example, you are writing a data sheet for a new product. You know that a number of competing companies make similar products, that some of these are as good as your company's, and that several are even better in a number of ways. Is it your responsibility to describe the ways in which the competitors' products are better than yours? Most people would say no, provided you have been honest in describing yours. A reasonable consumer can be assumed to understand that when different companies make similar products, each is a little bit different.

Keep one important point in mind, however. It is your responsibility to offer as much information as you can about anything that could affect the safety of the person using the product. If, for example, you know that the chainsaw you manufacture can cause serious injury when used improperly, you must do everything you can to explain how to use it the right way. The fact that a competitor's chainsaw incorporates some design innovations that make it inherently safer than yours does not excuse you from the duty to try to prevent injury to the user. You must explain—as effectively as you can—how to prevent injuries, even if in doing so you alert your reader to the inferiority of your design.

Modern trends in liability suits encourage complete honesty in communication. Juries want to see that companies have done everything possible to give their customers full and complete information on which to base their decisions. Honesty is the best policy, both ethically and financially.

Clarity

Each statement in the document should convey a single meaning that the reader can understand easily.

Unclear writing is expensive. Over a decade ago, a typical letter cost about $10 in labor and materials; the cost now is probably double that (Day 1983). But these dollar figures are misleadingly low because of the cooperative nature of most projects today. While an unclear document is being rewritten, a whole team of people can be waiting. Or, even worse, a team can start to work based on the information contained in an unclear document. Incorrect quantities of materials are purchased, construction begins in the wrong location, and so forth.

Unclear writing can also cause ethical problems. Ambiguous or confusing warnings on medication bottles and unclear instructions on how to operate equipment can cause sickness, injury, or even death. An unclear construction code can result in unsafe roads and buildings.

Accuracy

All the problems that can result from unclear writing can also be caused by inaccurate writing.

Accuracy is a simple concept in one sense: you must get your facts right. If you mean to write 2,000, don't write 20,000. If you want to refer to Figure 3–1, don't refer to Figure 1–3. Inaccuracies are at least annoying and confusing; they can also be dangerous, of course.

In another sense, however, accuracy is more complex. Workplace writing must be as objective and unbiased as you can make it. If your readers suspect you are slanting the information—by overstating or omitting a particular point—you will lose your credibility, and they might not believe anything you say.

Comprehensiveness

The document should include all the information the readers need (or at least cross reference other necessary documents). An effective document includes a background section if the audience includes readers unfamiliar with the project. An effective document also includes a clear description of the writer's methods, as well as a complete statement of the principal findings—the results and any conclusions and recommendations.

Comprehensiveness is crucial for two reasons. First, the people who will act on the document need a complete discussion in order to apply the information safely, effectively, and efficiently. Second, the document functions as the official record of the project from start to finish.

For example, a scientific article reporting on an experiment comparing the reaction of a virus to two different medications will be published only if the writer has explained his or her methods in detail; other scientists might want to replicate the experiment. Or consider a report recommending that a company enter a new product area. Before committing itself to the project, the company will want to study the recommendation carefully. The people charged with this responsibility need all the details. And if the company decides to go ahead with the project, the report will serve as the official documentation for the project. Months or years later, company officials will know where to turn to find out what was done and why.

Accessibility

An accessible document is structured so that readers can easily locate the information they seek. Most technical documents are made up of small, independent sections. Some readers are interested in only one or several of them; other readers might read most of them. But relatively few people will read the whole document from start to finish, like a novel.

Therefore, you should make your document easy to access by creating self-contained discussions and by using headings and lists (see Chapter 5) and, for reports, a detailed table of contents (see Chapter 15). A consistent page design (see Chapter 10) also helps readers find information.

Conciseness

For a document to be useful, people have to read it, and a short document is much more likely to be read than is a long document. Therefore, your writing should be as concise as you can make it without sacrificing the other criteria of effective writing.

One way to shorten a document is to get rid of the long words and phrases. Instead of writing, "The failure rate must be taken into consideration," just say, "The failure rate must be considered." Before writing, "The fact of the matter is that. . . ," ask yourself if the phrase says anything at all.

The real enemy of conciseness, however, isn't the individual word or phrase. Rather, it is the "at-least" principle discussed earlier in this chapter: the bad idea that long documents are better than short ones. When our bosses tell us that they want no more than 3 pages, we mistakenly assume that they really want at least 10, and 15 will show real effort. But your readers are just like you; they prefer short documents.

Correctness

Writing should observe the conventions of spelling, grammar, punctuation, mechanics, and usage. Some of the conventions are important in an obvious way: if you write "While feeding on the worms, the researchers captured the birds," you've got the researchers eating worms. Most of the conventions, however, are important because they make you look professional. If your document is full of careless writing errors, your readers will begin to doubt the accuracy of your technical information. Although some very bright people can't spell, most of them use a dictionary and a spell-checker and ask other people to help them.

Diplomacy

Effective writing is polite and gracious and avoids needless insults. If you are writing to a supplier who sent defective materials, nothing will be gained if you say, "Who do you think you're dealing with, a bunch of chumps? These microchips you sent us all failed the quality-control tests, and we're not going to pay for them." Instead, be precise and restrained: "In the shipment of 2,000 microchips (order 357-968), 14% failed quality-control tests. This failure rate violates our agreement, and we will not be able to authorize payment until the defective chips are replaced." Even when you have to adopt a firm tone in your writing, basic politeness is the best policy, for it encourages your reader to act professionally toward you in response.

Effective workplace writing is honest, clear, accurate, comprehensive, accessible, concise, correct, and diplomatic. Notice that I didn't mention stylistic individuality. Workplace writing is meant to get a job done, not to show off your personal style. The reader should not be aware of your presence. As a writer you create the document, so your influence permeates the document, but the reader should be thinking about the subject you are writing about, not about you.

Reference

Day, Y. L. 1983. The economics of writing. *IEEE Transactions on Professional Communication* 26, no. 1: 4–8.

Chapter 2

Before You Start Writing

This chapter presents some basic principles and planning activities that will save you time and make your document a lot more effective. First, I offer two tips about the overall process of writing: use short writing sessions, not long ones, and collaborate whenever you can. Next, I describe four steps that you should carry out to complete your planning: analyze your audience, define your purpose in writing, determine how your audience and purpose will affect the shape of the document, and, finally, make sure your boss agrees with what you plan to do.

Use Short Writing Sessions, Not Long Ones

When people apportion their time to work on a writing project, they often make a serious mistake: they set aside a big block of time so that they can concentrate fully. If they know they have about five hours to devote to an important memo, they set aside Thursday afternoon and part of Friday morning. In this way, they think, they can devote their full attention to the project and really make it good. This is a technique they picked up in school: for the physics test on Tuesday morning, start studying Monday at midnight, and brew a big pot of coffee.

It didn't work for the physics test, and it doesn't work for workplace writing. If you know you have only five hours to devote to the memo, it's much smarter to spend one hour a day for five days than to spend five hours on one day. Breaking up the time into little pieces is more effective for two reasons:

- It's more consistent with the way the mind works. Writing—or even trying to figure out what you want to write—is very hard work. Few people can maintain their sharpness for more than an hour or two without a break. When you keep the writing sessions relatively brief and spread them out over several days, you not only prevent the fatigue that leads to wasted effort, but you also give your mind a chance to work on the project while you take a break.

  Have you ever noticed that sometimes you get a really good idea about how to solve a problem in a document when you're not thinking about it? You're driving home, or taking a shower, or even sleeping, and all of a sudden the document falls into place. Obviously, you must have been thinking about it, or you wouldn't have gotten the idea. But you weren't consciously thinking about it. Your brain never shuts down, and sometimes it does its best work when you're not pestering it. Therefore, if you spread out the writing over several days, you increase the chance that your brain will figure out what to do while you're taking it easy.

- Breaking up the writing sessions into smaller periods helps you forget what you've written, and that makes you more like the reader of your document and less like the writer. If you spend a whole day working on the memo, you can get stuck in a rut. Not only can you waste long stretches of time producing nothing, but when you do write something you become too attached to it. Whatever you have written gets etched onto your brain, and you can't get a good perspective on it to see whether it makes sense. But if you work on the memo every day that week for an hour, you get a chance to forget it three or four times. Each morning, when you first reread what you produced the day before, you are seeing it fresh. In the first pass through yesterday's writing, you will notice all sorts of things you want to revise—things you never would have seen the day before even if you had spent another five hours at the desk.

Write Collaboratively

Everybody has a memory of a teacher telling them, "Do your own work!" I recommend that, whenever possible, you work with your neighbor.

The research suggests that people write collaboratively frequently in business and industry. According to one survey, some 87 percent of professionals reported that they write collaboratively at least some of the time (Ede and Lunsford 1990, 60). Many projects are just too big and complicated for one person; who has the time and all the skills to put together a full-size manual? But even if collaborating were not necessary, it still would be the smart way to make important documents. The principle is the same as in segmenting the writing sessions: if you can bring more people into the process, you increase your chances of seeing your document as your real readers will.

Collaboration takes many forms. There can be individuals or groups involved at the planning, drafting, and revising stages, and you will probably have to experiment to see which approach works best for you for a particular kind of document. In general, the larger or more important the document, the more useful it is to think of collaborating. A 200-page procedures manual almost certainly will call for collaboration during the planning and revising stages, and the drafting might be carried out by several people too. But a 2-page monthly status report might be planned and drafted by one person and then edited by another.

People have devised a number of collaborative techniques that work smoothly during the planning, drafting, and revising stages.

The Planning Stage

For getting started, it is best to gather several people in a room for about an hour. Then start generating ideas that might go into the document. A number of different techniques work well, including sketching, free writing, and brainstorming (these techniques are discussed in greater detail in Chapter 3). In the early stages, the group can go beyond generating ideas to planning how to organize the document and how to integrate the graphics.

The Drafting Stage

Sometimes one person takes over at this point and drafts the document; other times, the group remains intact. In these cases, after the group members establish a strategy for the document, they delegate tasks. For instance, first they make a brainstorming list, turn it into a structured outline, and settle on a style (regarding paragraph length, vocabulary, and many other tactical issues). They then decide who will do what. Sometimes the decisions are easy to make: the graphic artist does the graphics, the lawyer writes the legal section, and so forth. But sometimes the decisions are more difficult. Should the lawyer write the legal section, or should he or she meet with the technical writer, who will actually do the writing? Once these decisions are made, the different people go off and do their jobs. For long, complicated, or important writing tasks, the group chooses a leader who coordinates the different participants' tasks and sets up meetings to report on progress and solve any difficulties.

The Revising Stage

After the document is drafted and is ready to be revised, collaboration is crucial. If you wrote the document essentially by yourself, give a draft to a person—or several people—you trust. Ask them to jot down notes in the margins about anything they don't understand or they think might not be effective. If you wrote the document as part of a group, revision is often a group process, with each group member studying the draft separately and then getting together to work out a consensus. Regardless of whether the document is a single-author or multiple-author effort, keep one point in mind: any feedback you get from another person is extremely valuable. Even if you disagree with it, the criticism shows how at least one reader interpreted what you were trying to say. And that reader's misinterpretation of your draft might well indicate what will happen when the real reader sees it. A basic principle of workplace writing is that even if a misinterpretation isn't your fault, it's still your problem.

Analyze Your Audience

Whether you're working alone or collaboratively, there are certain steps you can take before you begin to write that will make the process less painful, more successful, and quicker in the end. The first and most basic of these is to define your audience.

You've got your data in front of you, and you might be tempted to start writing. Don't; instead, stop and think about your audience. Unless you know who will be reading what you have put down on paper—and why they will be reading it—you cannot be sure of what to say or how to say it. Think carefully about four aspects of your audience: their professional characteristics, their personal characteristics, their attitudes toward the subject, and their reasons for reading the document.

Professional Characteristics

What are their positions in the company? What are their areas of responsibility? How familiar are they likely to be with the subject you are writing about? Is their knowledge current and accurate, or will they require an update?

Personal Characteristics

Some readers like long sentences and paragraphs full of technical concepts; others hate them. Some readers are fully comfortable with computer jargon, such as WYSIWYG, whereas others won't have the slightest idea what you're talking about. Common sense will tell you more. A 22-year-old employee just out of college won't think like a 20-year veteran of the company. Someone whose interests may prosper from what you have to say will read your document enthusiastically; someone whose interests may suffer will read it with a special kind of vigilance.

Attitudes Toward the Subject

Will the readers be sympathetic to the subject and the approach? Or does their position in the company or their professional experience put them in the role of skeptic—or even adversary? If you are writing about government regulations, for example, are your readers likely to view them as unreasonable burdens or as necessary instruments intended to prevent abuses?

Reasons for Reading

Are the readers technical personnel who have to implement your recommendations or use your document as a working tool? Or are they managers who need only the bottom-line information to assist them in organizational planning or who simply need to know what you have been working on?

Keep in mind your profile of the readers. In a minute I'll discuss how to use this information.

Define Your Purpose

After you have defined your audience and its needs, ask yourself why you are writing.

Beware of vague answers: "I want to talk about the market for pen-based computers in the next five years." Don't think only about what you want to say; that's your subject, not your purpose. Consider what you want your document to accomplish. When your readers have finished reading, what do you want them to know or feel or believe or do?

The best way to determine your purpose is to force yourself to define it in one sentence:

I want to explain how the increased price of cadmium will affect our production costs.

I want to recommend that we investigate purchasing pen-based computers for the sales force.

I want to describe our level of success in meeting quality-control standards at the Altoona facility.

Notice in these examples that the infinitive verb (*to explain, to recommend, to describe*) does the real work. Your purpose will fall into one of two broad categories: to teach your readers something or to modify their attitudes toward a particular situation (and perhaps motivate them to take an action). Following are a few examples of "teaching verbs" and "attitude verbs" that you might use.

Teaching Verbs	*Attitude Verbs*
to explain	to request
to summarize	to requisition
to review	to authorize
to forecast	to propose
to describe	to recommend
to define	

Evaluate the Implications of Your Audience and Purpose

Understanding your audience and purpose has two important benefits:

- It gives you a sense of direction, a way to decide what belongs in the document.
- It helps you make the necessary strategic and tactical decisions, ranging from choosing an effective structure for the document to deciding about paragraph and sentence structure and word choice.

To show how the audience and purpose determine every aspect of the document, consider the following example.

As the head of a 10-person drafting department at your architectural design firm, you believe your department is losing some important contracts because you're using outdated hardware and software. You've mentioned the problem to your boss several times. One day she calls and asks you to spend a few days researching the available options and "write up your recommendations."

You start by sketching in some of the basic facts about your reader:

- She is a middle-level manager who cannot authorize large capital expenditures, but she can recommend—or not recommend—your ideas to her boss.

- She is not an expert on the kind of equipment you want to buy, but she isn't hostile to the idea of new high-technology equipment. She is unlikely to understand all the terminology you've picked up in your reading about CAD equipment.
- She knows that your department is productive and that you have never requested unnecessary or inappropriate purchases of equipment. On the other hand, the company did make an unwise purchase just 10 months ago, and nobody has forgotten it.
- She would prefer simply to attach a covering memo to your memo (if she approves it) and send it on to her boss rather than to rewrite it in her own words.

These facts tell you that you have to accommodate the needs of both your reader and her reader. You have to be direct, straightforward, and objective. You should include a brief summary for the convenience of your readers. Your explanation of technical concepts should be simple, and if you use technical terms at all, you should be sure to explain what they mean. You should stress the practical advantages of the new equipment.

Next, you define your purpose in writing. In this case, the purpose is clear: to persuade your readers that the new equipment is necessary and cost-effective so that they will authorize its purchase.

Finally, you examine your subject in light of your understanding of the audience and purpose. Because your readers will be particularly wary of large capital expenditures, you have to show very clearly that the type of equipment you want is necessary and that you have recommended the most effective and efficient models. To do this, your memo will have to answer the following questions that will be going through your readers' minds:

- How does your department currently perform its drafting functions?
- What is wrong with the current approach, or how would our operations be improved by new equipment?
- According to what criteria should the different options be evaluated?
- What is the best solution for the job?

- What have other purchasers of the equipment experienced?
- How much will it cost to purchase (or lease), to maintain, and to operate?
- Is the cost of the new equipment justified? At what point will it pay for itself in increased productivity or quality?
- How difficult would it be—how long would it take—to have the equipment in place and working?
- What is the learning curve? How long will the draftspersons need to become comfortable with the new equipment?

Carefully defining the writing situation is the first step in planning any writing assignment. If you know the persons(s) you are writing to and why you are writing, you can decide what to say and how to say it. The result will be a document that works.

Make Sure Your Boss Agrees with What You've Decided

You have a good understanding of your audience and purpose now, and a general outline is starting to take shape in your mind. But before you actually begin writing your outline or gathering the necessary data, it's a good idea to spend another 10 or 15 minutes making sure your boss agrees with what you've decided to do. After all, her request that you "write up your recommendations" about drafting equipment was quite vague. What she thinks she said and what you think she said might not be the same. The last thing you want to do is waste two or three days preparing a memo that will be rejected.

Submitting a statement of your understanding of the audience and purpose is a way to establish a kind of informal contract with your boss. Even though bosses can change their minds later, informal approval is better than no approval at all. Then, if your task is redefined, you at least have a piece of paper that explains what you have been doing with your time.

In this statement, clearly and briefly explain what you're trying to do. Figure 2–1 is an example of the statement you might submit to your boss about the drafting equipment.

Anita—
Does this seem like a good approach for the memo on drafting equipment?

The purchase price for the complete system for the whole department will be more than $30,000, so Grant would have to approve it. (I'll provide leasing costs as well for his information.) We want to make the case that the method we use now to produce the designs and blueprints is costly and time-consuming and that modern CAD equipment could decrease costs and increase output. The proposal should include detailed cost-benefit analyses and payoff period data.

I'll stop by tomorrow to get your reaction before I begin research.

Figure 2-1 Statement to a Supervisor

Once you have received your boss's aproval, you can feel reasonably confident in proceeding to write your document.

Reference

Ede, L., and A. Lunsford. 1990. *Singular texts/plural authors: Perspectives on collaborative writing.* Carbondale: Southern Illinois University Press.

Chapter 3

Understanding the Writing Process

You've done your planning: you've analyzed your audience and purpose, thought about their implications for the document, and made sure your boss agrees with your analysis of the situation. It's time now to begin the writing process: gathering your ideas, organizing them, drafting, and revising.

Generate Ideas to Include in the Document

Before you can start to write a document, you need to decide what information it will contain. Start by generating ideas—that is, by trying to think of the topics you want to discuss or the points you want to make. There are four popular methods of generating ideas: brainstorming, talking, free writing, and sketching.

Brainstorming

Brainstorming is the name for the process of quickly listing all the ideas that might be appropriate for the document. Brainstorming can be carried out alone or in a group, and the ideas can be recorded on any kind of writing surface, such as a computer screen, a piece of paper, or a whiteboard.

The essence of brainstorming is speed. A brainstorming session for a short report might take only 10 or 15 minutes. Write down ideas quickly, without trying to organize them and without pausing to decide whether they are appropriate. Brainstorming is the process of stockpiling raw materials that might be useful; it is not quality control. The ideas should be mere phrases, just long enough to remind you later of what you were thinking of. For instance, "need to update the parts catalog" is sufficient.

Although brainstorming alone is common for short, simple documents, collaborating works better for longer, more complicated ones. Collaborative brainstorming creates a synergistic effect: one person's idea sparks another person's, and that idea sparks another's. Four people in a room can produce a lot more ideas in an hour than any one of those people could in four hours. And because the ideas will represent a much broader perspective than any one person's, the finished document will more likely meet the audience's needs.

Talking

Simply talking with another person is an effective technique for generating ideas, especially when you get stuck. You can't figure out what you want to say or what should come first. Perhaps you're getting lost in a sea of details. Sit down with someone who is willing to ask you some questions. You start by describing, in one sentence, what you are writing about: "I want to propose that the Drafting Department be authorized to purchase new computers and software for CAD." The person then begins a conversation with you: "What's wrong with the equipment you have now?" Soon, a substantive discussion is underway, and many of your roadblocks will disappear as you explain what you are doing and why. If the problems don't disappear, at least you will have a better understanding of what they are.

Free Writing

This is an attempt to get beyond writer's block, the feeling of despair that sets in when you stare at a blank page or screen and don't know how to proceed. Free writing consists of writing nonstop for a given period of time, such as 10 or 15 minutes. What do you write? Anything that comes to mind about the subject. You don't pause to think about what comes first. Instead, you just make sentences. If you can't think of what to say, write about your inability to think of something. A free-writing session of 3 minutes might produce something like Figure 3–1:

> I need to get hold of some reviews of the best CAD software. The manufacturer's rep gave me some product sheets, but I know Grant is going to be looking for reviews. Maybe Sally can get hold of these for me. Also, I need to find out if any of the journals have done any head-to-head comparisons of the leading software. Maybe Bob over at Blair Consulting can help me out—he owes me one. I also need to find out if we can get a site license or at least a quantity discount on the software.

Figure 3–1 Free-Writing Passage

The advantage of free writing, like brainstorming, is that you don't worry about where an item will appear in the document—or even *whether* it will appear. Free writing is effective for people who are particularly anxious about their writing—the people who are so worried about their final draft that they find it hard to make a first draft. Once they do a few free-writing sessions, they relax and the ideas start to flow. Nothing generated in a free-writing session is likely to appear unchanged in the final document, but the technique is a simple and efficient way to put some ideas on the table.

Sketching

This is a particularly effective technique if you think visually rather than verbally. Instead of committing yourself to a linear structure, with one idea presented first and another presented second, sketching lets you present information in a nonlinear format. Because you aren't worrying about organization, you can concentrate on generating ideas.

One sketching technique is called *clustering.* Start by writing a phrase in the middle of a sheet of paper or on a chalkboard; then draw a circle around it. That phrase might be "proposal for new CAD software." What comes to mind immediately? "Problems with existing software." Write this phrase somewhere else on the page or the whiteboard, draw a circle around it, and connect it to the first phrase. This second phrase now looks like a satellite around the first phrase, the planet. As other elements of the main idea come to mind, add them in the same way. Don't forget to put satellites around the satellites. For example, "problems with existing software" can have satellites such as "too slow," and "3D too primitive." The advantage of planets and satellites is that you don't have to make any decisions about organization and sequence; you're just trying out ideas.

Figure 3–2 is an example of clustering from early in the idea-generating process.

A second sketching technique is called branching. Start with the key phrase at the top of the page: proposal for new CAD software. Underneath it you add the second-level ideas, each of which might have its own subordinate ideas. Soon you are branching out.

Figure 3–3 is an example of branching.

Organize Your Information

Eventually you have to organize the information by deciding how important each item is and by working out a sequence for the items. Some people are comfortable organizing their ideas after they have generated them using one of the techniques just described. Other people, however, prefer to put off the organizing until after they have drafted. It doesn't matter which approach you use.

If you want to start to organize at this point, you might have to overcome your bad feelings about outlines. Even though almost everyone who writes a lot uses some kind of outline, most people remember it as a classroom exercise—a set of Roman numerals, Arabic numerals, and capital letters—not as a useful tool. The for-

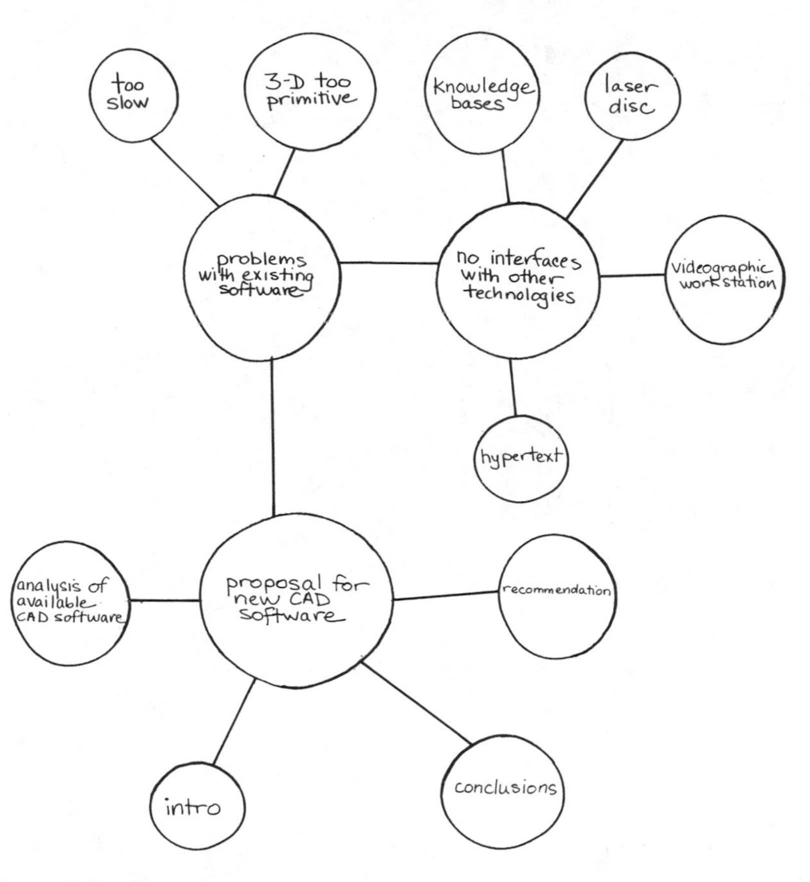

Figure 3–2 Clustering

mat is unimportant. Write the outline on the back of an envelope. Use phrases that mean something only to you. Do whatever you want, but force yourself to write some kind of outline. You'll save time in the end.

Organizing the information consists of three steps: stating your overall purpose, creating logical categories, and sequencing the ideas within the categories.

1. *State Your Purpose*
   Force yourself to write down your purpose at the top of the page: "To make the case that new CAD equipment will improve operations." If you don't constantly focus on your purpose as you write the outline, you won't be able to weed out items that don't belong.

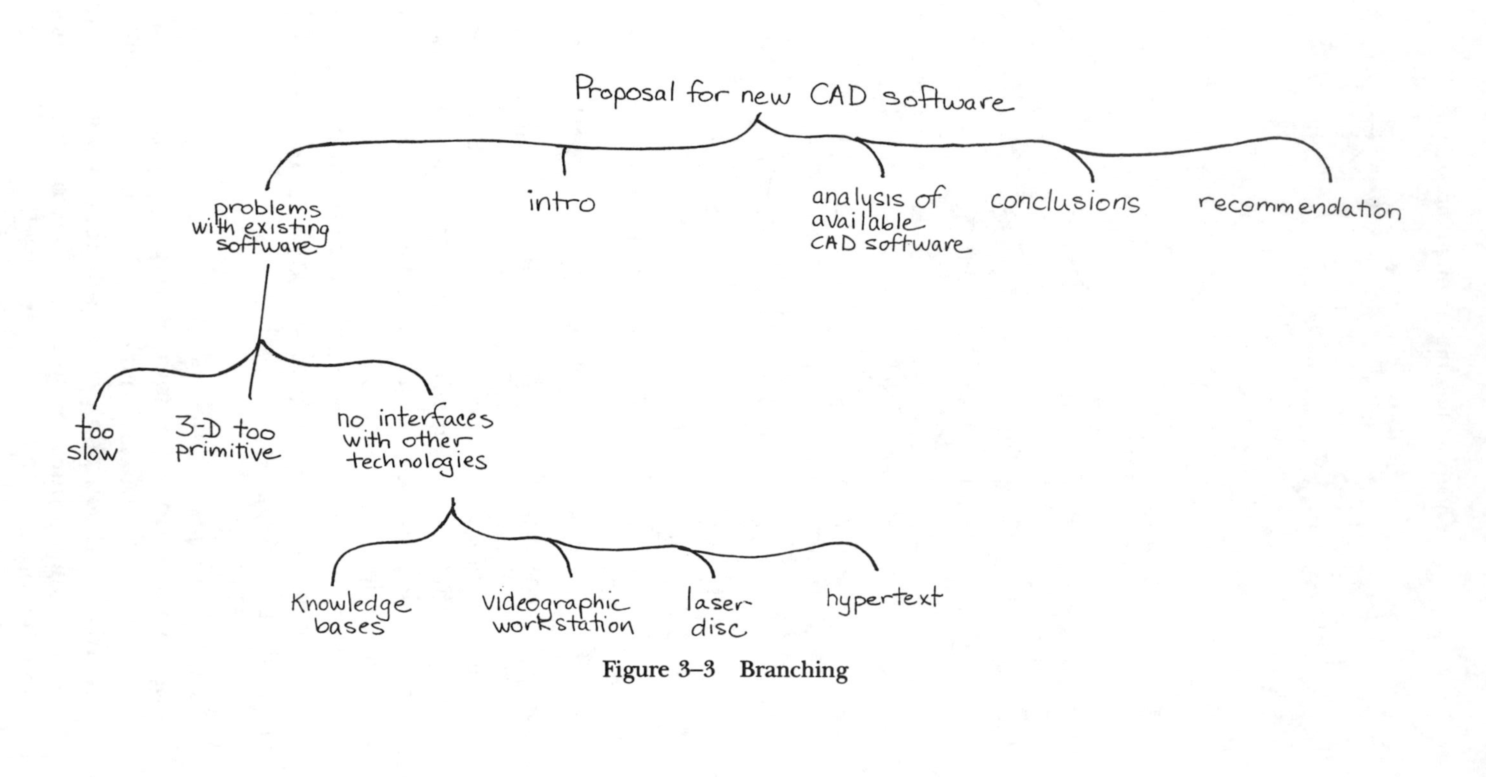

Figure 3–3 Branching

2. *Create Logical Categories*
   Draw lines between related items on the paper. In writing about the CAD equipment, for example, all the items about your department's current method of drafting would form one group. Another group would be the items about the cost of implementing the new system. Soon all the items will either be incorporated into groups or thrown out because they are irrelevant. Once you have some three to seven major groups, perform the same process on each individual group to create subgrouping.
3. *Sequence the Ideas*
   Arrange them in a logical order. What is a logical order? Any sequence that makes sense to you and addresses your audience's needs. Keep in mind that most complex documents use two, three, or even more different organizational patterns: one pattern, such as general-to-specific, for one level, and another pattern, such as chronological, for the second level. Consistency is a virtue, but don't bend your ideas to make them fit a pattern; rather, choose the pattern that best communicates your ideas.

Most documents can be structured according to one of eight basic patterns or some combination of them. The following discussion of each pattern will begin with a brief explanation, followed by a small portion of an outline showing how the pattern might look.

Chronological. The chronological pattern works effectively when you want to describe a process. If your readers have to follow your discussion in order to perform a task or just to understand how something happens, chronology is probably the most natural pattern.

> Purpose: to explain the steps in the volume visualization process
>
> —Steps in the Volume Visualization Process
>
> 1. Gather the data.
>   1.1 Arrange the data into slices.
>   1.2 Process the slices so that they cover a wide distribution of values, are high in contrast, and are noise-free.

2. Reconstruct the data.
    2.1 Construct new slices where necessary.
    2.2 Interpolate to estimate the missing values.
    2.3 Scan-convert the irregular grids onto Cartesian grids.
3. Map the data into geometric or display primitives.
4. Store or manipulate the primitives.

Spatial. The spatial pattern is effective in structuring a discussion of a physical object or scene.

Purpose: to describe the standard computer keyboard

3. the numeric keyboard
    3.1 the top row: the "soft" function keys
    3.2 the middle rows: the number keys
    3.3 the bottom row: the arithmetic keys

Classification. The classification pattern involves placing items into logical categories to highlight important characteristics.

Purpose: to describe the major categories of tracking systems

1. major categories of tracking systems
    1.1 magnetic systems
        1.1.1 operating principle
        1.1.2 strengths
        1.1.3 weaknesses
        1.1.4 major vendors
    1.2 acoustic systems
        1.2.1 operating principle (etc., as in 1.1)
    1.3 inertial systems
    1.4 mechanical systems
    1.5 optical systems

Partition. The partition pattern involves taking a single physical entity and breaking it down into its logical components.

Purpose: to describe an amoeba

1. parts of an amoeba
    1.1 nucleus

1.2 ectoplasm
1.3 endoplasm
1.4 contractile vacuole
1.5 food vacuole
1.6 pseudopods

General to Specific. The general-to-specific pattern is useful in helping your readers gain an overview of a complex situation before you present the details.

Purpose: to explain how to apply K14 Plastic Coating

1. principles of K14 application
   1.1 how K14 bonds
   1.2 K14 on different floorings
   1.3 how K14 dries on the floor
   1.4 mixing the proper amount of K14
2. how to apply K14
   2.1 determining where to begin
   2.2 coordinating several workers
   2.3 wetting the applicators
   2.4 applying K14
   2.5 stopping for the day

Notice how this writer uses a chronological pattern for Section 2 of her outline.

More Important to Less Important. The more-important-to-less-important pattern is useful when you want to describe a number of points that don't lend themselves well to a chronological or spatial pattern. Readers generally want to learn the most important point first.

Purpose: to convey the test results on the Altair PC

1. introduction
   1.2 customer's statement of PC irregularities
   1.2 explanation of test procedures
2. problem areas revealed by tests
   2.1 faulty power supply
   2.2 loose network card

3. components that tested satisfactorily
   3.1 hard drive
   3.2 motherboard
   3.3 expansion boards

Problem-Methods-Solution. The problem-methods-solution pattern is effective because it conforms to the reader's expectations about the problem-solving process.

Purpose: to explain how we solved the problem of power outages disabling our computer operations

1. Problem: Power outages disable the minicomputer.
   1.1 frequency: 4 times/year
   1.2 duration: average 3 hours
   1.3 functions affected
      1.3.1 dispatching
      1.3.2 maintenance
      1.3.3 payroll
      1.3.4 accounting
2. Methods: Study alternative power systems.
   2.1 interactive diesel generator/UPS combination
   2.2 LP-powered engine generator
3. Solution: Purchase LP-powered engine-generator.
   3.1 Microprocessor control interfaces with UPS battery bank for automatic turn on and turn off.
   3.2 Generator cycles automatically every 14 days to ensure proper operation.
   3.3 Microprocessor-controlled maintenance log monitors fuel and oil level and performs periodic maintenance.

Cause-Effect. The cause-effect pattern is useful when you want to discuss your subject as a phenomenon that was caused by several factors or as a phenomenon that caused (or might cause) other factors. You can reason from cause to effect (as in the following example) or from effect to cause.

Purpose: to explain the benefits of installing a LAN

1. benefits of installing a LAN

1.1 organizational benefits
  1.1.1 coordination
  1.1.2 centralization
1.2 technical benefits
  1.2.1 increased speed of data processing
  1.2.2 sharing of data and software
  1.2.3 increased physical mobility within the plant
1.3 financial benefits
  1.3.1 inexpensive access to mainframe
  1.3.2 decreased capital costs for new workstations and software
  1.3.3 improved control of finances

The writer here is arguing that installing a LAN (cause) would have three major categories of benefits (effects).

After you've written the basic outline, let it sit for a while. Come back to it in a few hours. Check it for logic, completeness, accuracy, and emphasis, and make sure it is consistent with your analysis of audience and purpose. Once you've started to write the document, you'll be concentrating on the individual sections, and you'll find it harder to focus on the big picture.

Following is a portion of the outline on CAD software.

Purpose: to recommend that we purchase new CAD software

1. introduction
2. problems with existing software
  2.1 too slow
  2.2 3D too primitive
  2.3 no interfaces with other technologies
    2.3.1 knowledge bases
    2.3.2 laser disc
    2.3.3 videographic workstations
    2.3.4 hypertext
3. analysis of available CAD software
4. conclusions
5. recommendations

Notice that at this stage in the prewriting process, the writer has thought in detail only about section 2: the problems with the existing software. Yet he has already decided on an overall plan for the report: the problem-methods-solution organization. Within section 2, he will want to use another organization, probably more important to less important. But he doesn't yet know which of his three subsets (2.1, 2.2, or 2.3) will be most important; that decision will probably wait until he has drafted those sections.

Write the Draft Quickly

There is no single way to write the first draft, but many successful writers have found that the key is to relax and write as fast as they can without lapsing into gibberish. Some writers actually force themselves to draft for a specified period, such as an hour, without stopping. Writing the draft is closer in spirit to brainstorming than to outlining: you are just trying to turn your outline into sentences and paragraphs quickly.

Your goal is to get beyond the writer's block. Start to write; don't worry about individual words and sentences. Once you get rolling, you'll be able to see how well your outline works. The virtue of not stopping to worry about your writing is that you don't lose your concentration. Some writers are so insistent on keeping the rough draft flowing that they don't even stop when they can't figure out an item on their outline; they just pick up a new piece of paper or move the cursor and start with the next item that does make sense.

One more piece of advice about drafting: sometimes the document is so long that you cannot draft the whole thing at one sitting. When you have to stop a drafting session, *don't* stop at a logical juncture, such as a major section. Instead, go past that major juncture—at least two or three sentences into the next section—and then stop. This way, when you start drafting again you will be in the middle of an idea, and your thoughts will flow better. You will feel less as if you are starting from the beginning.

Spend Your Time Revising

For many writers, revising is the most arduous aspect of writing, but it is absolutely critical. An inexperienced writer, especially one who has crafted the first draft carefully, revises ineffectively. Because the first draft has actually been mulled over many times during its creation, the writer thinks it sounds pretty good—and that's the problem. The writer cannot get any distance from the writing and therefore cannot see it the way the real readers will.

Revising effectively involves four steps: letting the document sit, reading it aloud, getting help from someone else, and looking for particular problems.

1. *Let It Sit*
   Leave the document alone for as long as you can—at least overnight if possible. Forget it.
2. *Read It Aloud*
   This will do two things for you. One, it will slow you down, so you'll have a better chance of seeing things that should be changed. (If you watch a person revising a document silently, you will notice that the process picks up speed with each page. Pretty soon, the pages are flipping past every fifteen seconds; the final pages don't get much attention.) Two, reading it aloud will give you a chance to hear it, that is, to hear the flow of the sentences. Often, when you proofread silently, you don't notice that words are missing or that there are awkward phrases. Listening to the document helps you catch many of these problems. You may be wondering if you will feel foolish reading aloud. You will; join the club.
3. *Get Help from Someone Else*
   The more suggestions you can get from readers who are similar to your readers, the better the document will be.
4. *Look for Your Particular Problems*
   Inexperienced writers don't realize that each person makes the same mistakes over and over again. A good writer has just a few problems; a poor writer has many. But they tend to be the same ones from one document to the next. Therefore, it is necessary to figure out what your weaknesses are. Maybe someone around the office can help; or maybe the company sponsors a writing course or pays tuition for a night school or college course. Once you get a sense that you have a particular grammar problem, for example, you have to figure out how to locate it and remove it. Most problems have fingerprints that you can learn to identify. For instance, nominalizations (see Chapter 6)

often end in *-tion* or *-sis* or *-ment*, and they often are followed by the word *of*. Once you learn that, it's fairly easy to find them either by hand or using the search function on your word processor. (The word processor and style programs are discussed in Chapter 4.)

But how do you actually revise a document? Start by realizing that you cannot look through the document once and hope to see everything that needs to be changed. Revising requires going through it carefully numerous times, each time looking for one specific thing. Checklists, such as those in Appendix A, can be very helpful because they force you to focus on a specific point each pass through the draft. It's a good idea to see if your company has a checklist; if it doesn't, modify the ones in Appendix A to meet the needs of your subject, type of document, and audience.

Be prepared to spend well over half of your total time revising. By contrast, drafting is a snap; it should take only about 10 percent of the total time. Check the document for the following kinds of problems:

- *Comprehensiveness.* Have you addressed everything that is on the outline? Sometimes in drafting quickly you skip a point and forget to pick it up later. Even if everything is addressed, have you gone into sufficient detail to make your message clear?
- *Accuracy.* When you draft quickly, you can make errors of fact or leave out critical data. Now is the time to go over everything to make sure it's accurate.
- *Organization.* Have you explained the organization to your readers? Is it clear and logical? Does it flow from one point smoothly to the next?
- *Emphasis.* There should be a rough equivalence in length between a subpoint in one section and the same level of subpoint in another location. If one is three lines and the other is three pages, you might want to rethink the structure.
- *Paragraphing.* Does each paragraph have a clear topic sentence that forecasts or summarizes the information that follows it? Is the support logically organized and developed? Are there sufficient coherence devices, such as repetition of key terms, transitional words and phrases, and demonstrative pronouns with nouns?

- *Style.* Check for sentence variety, grammar and punctuation mistakes, and awkward constructions. It's a good idea to use a checklist (see Appendix A) to help you look for problems.
- *Spelling.* Nothing else makes you look unprofessional quite as fast as a few strategically placed spelling errors. Writers come in two varieties: those who know when they don't know how to spell a word and those who don't. Unfortunately, the second group is much larger. Spelling is one area where you need other people's help. And, of course, a computer doesn't hurt, either. The next chapter discusses how to use spell checkers.

The writing process described here might sound time consuming, and it is. But with a little practice you'll be able to generate your ideas fairly fast. Drafting goes quickly. The time-consuming part is the revising. Yet there is no getting around it. If you want the document to reflect your best thinking, you have to use a method that exploits how the brain works and not ask it to do things it cannot do, such as instantly create a full-blown outline out of nothing or revise for twelve things at once. The process described here is, for most people, the quickest way to high-quality writing.

Chapter 4

Using the Computer to Improve Your Writing

The personal computer has existed for over a decade, and it keeps getting smaller, more powerful, and less expensive. As an example of technology, nobody would deny that it is remarkable. But this chapter discusses the computer as a writing tool. How good is it? And what are the best ways to take advantage of its power?

In this chapter I don't talk about different brands of software or hardware. You can benefit from an Apple Macintosh or a PC, from WordPerfect or Microsoft Word. So many excellent products are on the market now that it really doesn't matter which ones you use. The important task is to learn how to take advantage of the basic features of all professional-level machines and software.

This chapter traces the path followed in the previous chapter, focusing on some basic ways to use the computer as a tool during the different stages of the writing process.

Understand What the Computer Can—and Cannot—Do for You

The computer is a terrific tool for writers. It can increase your writing productivity significantly. It can help you eliminate misspellings and grammar mistakes. It can help you produce professional-looking documents in much less time than you ever thought possible.

So how come everyone with a computer isn't a terrific writer? How come you still don't look forward to writing? The answer is simple. The computer can do only a few things well, and they're not the most important things. The computer can't think, and it certainly can't write. It specializes in minor housekeeping chores, such as moving things from one place in a document to another or changing all the uppercase letters to lowercase. It is patient, obedient, and very fast. Like a faithful dog that loves to fetch a stick for hours on end, the computer can find any set of letters you ask it to. The problem is, however, that you have to know what letters you want it to find. And once you've got them, you have to know what to do with them.

Expecting a computer to make you an excellent writer is a little like expecting to become a master cabinet maker by purchasing a basement full of sophisticated woodworking tools. If you know how to write, the computer will help, and if you know how to make cabinets, the sophisticated woodworking tools will help. But keep in mind that people made excellent documents and beautiful cabinets before the sophisticated tools were invented.

Use the Computer When You're Just Getting Started on a Document

Perhaps the most effective way to brainstorm is in front of a computer. Because even slow typists can work quickly on a word processor, you can create a lengthy brainstorming list fast. With the text-moving functions of a computer, you can easily rearrange items to turn a brainstorming list into an outline. Without the drudgery of rewriting or retyping, you are more likely to try out alternative groupings as you classify items. The same holds true when you sequence the groups. Of course, you can cut and paste on paper, but the process is much more cumbersome.

You can easily make copies for everyone on your team (if you are collaborating), and everything will be easy to read. Nobody has to copy over lists from a chalkboard. Other kinds of idea-generating techniques, such as sketching, can be done effectively with basic graphics software; again you have the advantage of hard copy that you can duplicate easily.

Use the Computer When You're Drafting

The computer is a useful tool during the drafting stage because it encourages quick drafting. Knowing how easy it is to move information gives you the freedom to begin writing wherever you want, rather than having to start at the beginning, which is the most difficult place to start. You don't have to worry about where you will ultimately put the discussion. And when you have trouble thinking of what to write, you can just skip a few lines and go to the next point on your outline. Rearranging the text later is simple.

The computer is a highly effective drafting tool for six reasons:

- *You can concentrate on your message.* Drafting on a computer is fast and easy because you don't worry about the quality of the writing or about typographical errors. You can give your full attention to what you are trying to say, because you know how simple it will be later to make major and minor revisions.
- *You can turn your outline into a draft.* Working right on your outline helps you stay on track. You just put the cursor at some point on the outline where you feel comfortable and then start to draft. As you write, the outline below will scroll down. You don't have a separate outline and a separate draft; the outline becomes the draft.
- *You can make a longer draft.* The research on computing shows clearly (Hawisher 1989) that writers produce longer texts when working on a computer than when writing by hand. Producing a lot of writing quickly is the goal of drafting: you want to have material to revise later. Because computers are relatively quiet and easy to type on, you can generate much more writing in a

given period of time. You don't have the physical effort of writing manually, and you don't have to return the carriage at the end of the line as on a typewriter.

- *You can draft without stopping to revise.* Just turn the contrast knob on the monitor to darken the screen. This technique, called invisible writing, encourages you to close your eyes or look at hard-copy notes or at the keyboard. As a result, you keep typing. Because you can't stop and revise sentences, you will draft faster. In addition, your document is likely to be more coherent, because in drafting quickly you will be less likely to go off on tangents or forget the point you are trying to develop. (Some writers are afraid of invisible writing; they're not sure any words are actually making it to the screen. If invisible writing is not for you, try barely-visible writing: adjust the contrast just enough so that you can see something happening on the screen but you cannot actually decipher the words.)
- *You can abbreviate with the search-and-replace function.* This function lets you find any character, phrase, or word and replace it with any other writing. For instance, if you have to use the phrase *nondestructive testing techniques* often in your document, you can simply type *non** each time. Then, when revising, you tell the computer to change every *non** to *non-destructive testing techniques.* (You don't want to type just *non* as your abbreviation, because when you search-and-replace it, you'll also pick up every word that has *non* in it, such as *anonymous,* which will get turned into *anondestructive testing techniquesymous.*) This technique not only saves time as you draft; it also reduces the chances of misspelling, for you have to spell the phrase correctly only once.
- *You can collaborate more conveniently.* If you are collaborating and your computers are linked on a network, groupware (software that facilitates collaborative writing) makes it easier and faster to merge different people's text and graphics into one document.

Use the Computer When You're Revising

Computers make every kind of revising easier. One obvious advantage of revising with the computer is that your writing is legible, so you don't have to try to figure out your own handwriting. In addition, the neatly typed text gives you a more objective perspective on your work. You are seeing it as others will.

If the typographical errors are distracting, use the add-and-delete function to fix them so that you can concentrate on more-important changes. You can make major revisions to the structure and organization of the document easily using the block-move function. This function lets you try out different versions of the document without having to cut and paste pieces of paper. The copy function lets you copy text, anything from a single letter to an extended passage, without disturbing the original version. With the copy function, you can create two different versions of the document at the same time and then decide which one works better.

Most word-processing programs also contain a feature called redlining, which lets you track revisions. You redline the revisions, which makes it easier to identify them. Then, you can either keep the file for a permanent record of the revisions you have made, or you can remove the redlining with a few quick keystrokes and print the new document.

Although computers can help you do much of the work involved in revision, they cannot replace a careful reading by another person. Revise your document yourself; then get help from someone you trust.

Use Specialized Software

In the discussion so far, the word *computer* has been used to refer to a personal computer using a standard word-processing package. But there are three other kinds of software available that can help you identify problem areas that need to be fixed: spell checkers, thesaurus programs, and style programs.

Spell Checkers

A *spell checker* compares each word that you have typed against a dictionary, usually of more than 100,000 words. The program checks 5,000 to 10,000 words per minute, alerting you when it sees one that isn't in its dictionary. Although that word might, in fact, be misspelled, it might be a correctly spelled word that isn't in the dictionary. If it is a word the computer doesn't recognize, you can add it to the dictionary, so the computer will recognize it in the future.

When the computer thinks you have misspelled a word, it offers a list of suggestions—really just educated guesses about what word you're trying to write. Sometimes your spelling stumps the computer, and you have to look up the word in the dictionary and then change the spelling in your text. Without the spell checker to point out the error, you might not have known the word was misspelled.

One limitation of any spelling checker is that it doesn't know whether you have used the right word; it only knows whether the word you have used is in its dictionary. For instance, the computer would see no problem even if you typed "Wee knead two bye a gnu won bee cause the auld won baroque." Therefore, you still have to proofread carefully for spelling errors. The computer is content to let you use the wrong word—as long as you spell it right.

Thesaurus Programs

A related program is a *thesaurus,* which lists synonyms and near-synonyms for many common words. A thesaurus program has the same strengths and weaknesses as a printed thesaurus has: if you can't quite think of the word or phrase you want, the thesaurus will help you remember it. But don't forget that the words listed do not have exactly the same meaning as the key term. Unless you are aware of the shades of difference, you might be tempted to substitute an inappropriate word. For example, the word *fame* is followed in *Roget's College Thesaurus* by the word *notoriety.* If you don't know the difference in connotation between *famous* and *notorious,* you could embarrass yourself badly.

Style Programs

A *style program* performs several analyses of your writing; it computes such factors as sentence length and the number of passive voice constructions and expletives (*it is . . . , there is . . . , there are . . .*). Many style programs point out your use of abstract words and suggest more specific ones. Several identify sexist terms and provide nonsexist alternatives. A number warn you when you use a lot of prepositional phrases. Many point out fancy words, such as *utilize,* and suggest substitutes, such as *use.* And finally, a number apply readability formulas, measures of how easy the text is to read. (Readability formulas are discussed in Chapter 7.)

Keep in mind the limitations of style programs. Although they can find all your uses of expletives, they cannot help you determine which uses are appropriate in your document and which aren't. The research on style programs suggests that they are most valuable for experienced writers and least valuable for inexperienced writers (Crew 1988). Experienced writers are most willing to interpret the suggestions made by the program and best able to ignore the inappropriate or irrelevant advice. Less-confident writers, on the other hand, tend to get bullied by the software. For instance, several programs will flag your use of the word *nature* and tell you that it is a vague word. The software is trying to tell you to revise a sentence such as "The proposal was of an inappropriate nature" to read, "The proposal was inappropriate." But if you are using the word *nature* in a different way—as in "human's relationship with nature"—the advice could be misleading.

One alternative to a style program is the search function, which can perform many of the same functions. For instance, if you know that you overuse expletives, you can search for words such as *is, are, were, it,* and *there.* Some of the instances of these words, of course, will not be expletives, but you can revise those that are inappropriate.

References

Crew, Louie. 1980. The style-checker as tonic, not tranquilizer. *Journal of Advanced Composition* 8, no. 12:66–70.

Hawisher, G. E. 1989. Research and recommendations for computers and composition. In *Critical Perspectives on Computers and Composition Instruction,* edited by G. E. Hawisher and C. L. Selfe, 44-69. New York: Teachers College.

Chapter 5

Improving the Coherence of Your Writing

Because most workplace writing consists of technical information that your readers do not know as well as you do, they need all the help you can give them. Perhaps your most difficult challenge is to make the writing coherent. Coherence refers to the way the writing hangs together, the ease with which the readers understand the transition from one idea to the next.

This chapter discusses a number of techniques for ensuring that readers follow your argument. The basic principle behind all the techniques discussed here is that you have to tell your readers where you will be leading them before you start; if they understand what you plan to do and why, they will be better able to concentrate on the information you provide.

The chapter covers the structuring units critical to coherence: titles, headings, lists, introductions, and conclusions.

Write Informative Titles and Headings

Research has shown (see, for example, Huckin 1983) that titles and headings are critically important in orienting the reader, yet few writers devote much thought to them. I once worked for a contracting company building a nuclear power station. In compliance with NRC regulations, tons of paperwork were produced each month. Still, every one of the thousands of memos they wrote bore the same title in the subject heading: Limerick Generating Station, Unit 1. Not particularly informative.

A good title should clearly identify two factors about the document: its subject and its purpose. Here are a few examples:

Choosing a Laptop: A Recommendation

An Analysis of the Geffers 302 Packager

The Greenhouse Effect by the Year 2000: A Forecast

A Summary of the Research on the Health Effects of Secondhand Smoke

Notice that each title includes the subject—the technical topic, such as "choosing a laptop"—and the purpose, such as "recommendation." For more information about analyzing your purpose, see Chapter 2.

Headings sometimes share the same responsibility of indicating the subject and purpose, but since the reader is already into the document and presumably is following its logical development, the purpose is often clear or implicit and need not be stated.

Like titles, headings are important in helping your readers understand the direction you will follow, but they have an additional job. Because most workplace writing consists of self-contained units of information intended for different readers with different needs, abilities, and interests, headings also help them relocate the information they want when they come back to the document later. They don't have to read the whole thing again; they can just scan the document until they see the appropriate heading on the page. In addition, because headings appear in the table of contents, readers rely on them for help in locating a specific discussion in the document.

As you check each of your titles and headings, ask yourself three questions:

- *Is it sufficiently precise?* You want to help your reader understand where you are going in the text that follows.

Imprecise: Enhancing Production

Precise: Three Techniques for Enhancing Production in the Zuma Copper Mines

Could the title or heading apply equally well to any other document or discussion? If it could, you need to be more precise. For instance, if your title is "The Geffers 302 Packager," your document could be an analysis of the system, but it also could be a recommendation to buy it, or any of a number of different kinds of manuals.

- *Is it easy to read and understand?* Avoid the noun strings that seem to go on forever; they can be difficult to follow (Gleitman and Gleitman 1970).

Unclear: Shipboard Corrosion-Prevention Techniques Task Force Meeting Minutes

Clear: Meeting Minutes for the Task Force on Shipboard Corrosion-Prevention Techniques

- *Is it structured appropriately for the subject and the audience?* The "how to" form is effective for instructions:

How to Attach the Optional Paper Tray to the Printer
How to Apply for NIH Research Grants

Questions work well for less-knowledgeable readers:

What Are the Advantages of Switching to On-Line Documentation for Our Service Manuals?
Why Expand Our Operations in the Southwest Sector? A Plan for the Next Fiscal Year

The *-ing* verb clearly suggests a process:

Adding Additional Parking Capacity to the Millersville Plant: Third Quarter Progress
Computing Wind Shear: Questions and Answers for Commercial Aviators

Information on how to design titles and headings appears in Chapter 10.

Use Lists to Communicate Parallel Information

Many sentences in workplace writing are long and complicated:

> We recommend that Operations bid on this facility because it contains the latest production technology, it is fully computer controlled, and it contains ample space for processing, storage, and future expansion, both inside and outside.

Readers might find this a difficult sentence to understand and remember because they cannot concentrate fully; they have to worry about remembering all the information that comes after "because."

But when the information is presented as a vertical list, it is much easier to follow:

> We recommend that Operations bid on this facility for three reasons:
>
> 1. It contains the latest production technology.
> 2. It is fully computer controlled.
> 3. It contains ample space for processing, storage, and future expansion, both inside and outside.

Presented as a list, the location of the words on the page reinforces the meaning. Readers can *see* that the sentence contains three items in a list; this visual reinforcement enables them to concentrate better on the information. Readers are also assisted in that the three items begin at the same left margin, again emphasizing their parallelism.

Vertical lists like these are appropriate in almost all kinds of workplace writing, but if for some reason you have to arrange the information as a traditional sentence, use a horizontal format:

> We recommend that Operations bid on this facility for three reasons: (1) it contains the latest production technology, (2) it is fully computer controlled, and (3) it contains ample space for processing, storage, and future expansion, both inside and outside.

Notice in both the vertical and horizontal list that the lead-in indicates the number of items contained in the list: "three reasons." This strategy helps prepare readers for the size of the list, enabling them to focus on the information itself. Another reason to indicate the number of items is that sometimes, especially in longer lists, the writer or typist inadvertently omits an item; stating the number helps everyone see if an item is missing.

Often, bullets are more appropriate than numbers in vertical lists. Numbers work well when you want to suggest sequencing or priority (the number 1 item being the first in order or the most important), but use bullets otherwise, particularly if people are listed:

> The following three officers will be present at the meeting:
>
> - Helen Cartwell, President
> - Stuart King, Vice President
> - Chaman Jetra, Recording Secretary

The horizontal version of this list would look like this:

> The following three officers will be present at the meeting: Helen Cartwell, President; Stuart King, Vice President; and Chaman Jetra, Recording Secretary.

Notice how the semicolon acts as a supercomma.

For information on how to design lists, see Chapter 10.

Use Introductions to Forecast Discussions

Writers sometimes forget that the reader doesn't know the subject as well as they do. As a result, they jump right into the discussion without adequately explaining what information they are going to present, how they are going to present it, and why they are going to present it that way. Explaining these things is the job of an introduction. (A document can have one introduction—right at the start—or a separate one at the beginning of each major section.)

An effective introduction answers six critical questions for readers:

- *What is the subject?* Unless your readers know the answer to this one, you won't get anywhere. Answer this question directly, even if you suspect they already know.

> The subject of this report is the inability of the scientific community to reach consensus on the question of scientific fraud. Is its incidence rising? Do we have adequate measures for detecting fraud before publication? Is scientific fraud simply an ethical lapse on the part of the researcher, or does the system of rewarding research and publication actually encourage it?

- *What is the purpose of the discussion?* Just as readers need to understand what you are writing about, they need to know what you intend to do in the discussion.

> This report has two main purposes: to summarize the recent trends in manufactured housing and to recommend the most fruitful areas for future R&D and marketing for the industry.

- *What is the background of the subject?* The background is the information readers need to understand the discussion. Today, with the wide distribution of documents through photocopying and e-mail, the background is important because many of your readers might not be up-to-date on the subject.

> The background on this research project consists of two major factors. First, since the 1991 introduction in Europe of a highly effective morning-after abortion pill, pharmaceutical companies in the United States have cautiously explored the scientific and legal questions surrounding manufacture of a similar product. Second, the increasingly hostile political climate regarding traditional abortion has spurred interest in techniques that return more autonomy to the woman. For additional information on these two factors, see Appendix C, page 19.

Notice how this writer effectively sketches in the background and then cross references the reader to another location in the document for a fuller discussion.

- *What is the scope of the discussion?* The scope of a discussion is its territory: what is included and what is excluded. Following is a portion from the scope statement in the introduction of a manual:

> Part I of the manual introduces the new phone system, including information on how to operate the basic features including call conferencing, call forwarding, and automatic dialing. Note, however, that voice mail is not discussed here; it is treated in a separate pamphlet, "Understanding Your New Voice Mail System," which will be distributed next week.

- *What is the organization of the discussion?* You want your readers to know the organization of the discussion so that they can concentrate on the information without worrying about what will come next. The following statement is from the introduction to a letter:

> The first section of this letter describes the background of the legislation on nuclear waste dumping, because it is necessary to understand how the legislation took shape in order to appreciate how the subcommittee reached its decision. Then, the letter discusses the majority opinion and, finally, the minority opinion.

- *What are the key terms that will be used in the discussion?* If you are going to use one or several key terms throughout the discussion, the introduction is a logical place to define them.

> At Patrick Supplies, *flextime* is the policy that enables workers to choose their starting and stopping times, within certain guidelines. Flextime has two main purposes: to lengthen the period during the day when our customers and suppliers from different time zones can reach us, and to accommodate our employees' other commitments, such as dependent care.

Naturally, every introduction is different. Sometimes you will not have to address all these questions, and sometimes you can answer several of them in the same sentence. But the important point is that you must try to help your readers understand what you plan to do so that they can devote their full attention to the information you provide.

Use Conclusions to Complete Discussions

The word *conclusion* has two different meanings in writing. One meaning refers to the inferences drawn from technical data. If, for instance, federal regulations stipulate that the emissions of a certain toxin in the wastewater be less than one part per billion, but you are producing four parts, the conclusion would be that you are out of compliance with that regulation. This kind of conclusion is discussed in Chapter 18.

The other meaning of *conclusion* is the last part of a document or a section of a document. In this chapter we are talking about the second sense of the word.

Although some documents, such as parts catalogs, do not generally have conclusions, most do. When crafting a conclusion, make sure you answer four questions that might be on a reader's mind:

- *What are the main points established in the document?* After reading a document of more than a few pages, a reader is likely to forget some of the material, especially from the beginning. Therefore, it is a good idea to summarize the important ideas in a paragraph.

> Our analysis yielded two main conclusions. First, common nondestructive testing methods, such as infrared photography and liquid crystals, would be ineffective, despite their accuracy, because they are not sensitive to particular materials. Second, ultrasonics appears to be the most promising method, with the point-contact mode preferable to the immersion mode because it is more convenient and less expensive.

- *What should be done next?* Even though one project has ended, you might want to offer recommendations on the course of future work.

> In this chapter we have focused on those aspects of specification writing that are most commonly cited as difficult by new writers. We have not attempted a detailed discussion of all the aspects of specification writing. Therefore, we recommend strongly that you study Chapter 7 (on the technical section of the specification), Chapter 8 (on the general conditions of the specification), and Chapter 9 (on the bidding documents) before proceeding.

- *How can the reader find out more information?* Often the most appropriate way to conclude a technical document is to help your readers understand how to get more information. Sometimes, this is a sales message; sometimes, it is not.

> If you wish to be placed on the mailing list for future communications on the FGG project, please put an X in the box below and return this copy to my office. Thank you.

- *How can we help you in the future?* Often you conclude with an offer to provide future services.

> Here at Winwood Design we pride ourselves on our 38 years of offering the finest professional services to the Trenton area business community. We appreciate the opportunity to have served you, and we hope that your trust in us has been rewarded. If the need for our services arises again, please do not hesitate to let us know.

References

Gleitman, L., and H. Gleitman. 1970. *Phrase and paraphrase.* New York: W. W. Norton.

Huckin, T. N. 1983. A cognitive approach to readability. In *New essays in technical and scientific communication: Research, theory, practice,* edited by P. V. Anderson, R. J. Brockmann, and C. R. Miller, 90–108. Farmingdale, N.Y.: Baywood.

Chapter 6

Writing Better Paragraphs

Chapter 5 covered some of the basic structural elements that help you make your writing coherent: titles, headings, lists, introductions, and conclusions. The present chapter discusses another critical aspect of coherence: paragraphing. Like a full document or an extended passage, an effective paragraph has a structure: it begins with a topic sentence to orient the reader and set the direction, it supports its main point logically, and it contains words and phrases to help the reader make the transition from one sentence to the next.

Begin with a Clear Topic Sentence

As you may remember from school, a paragraph consists of one or more sentences that make up a main idea. Because the paragraph is the major unit of writing for expressing conceptual ideas, coherence is always a concern. The logic of effective paragraphing is simple: you state your idea and then explain it or defend it. Most of the difficulty in writing paragraphs would disappear if writers just began with the most important thought and let the details follow.

The major idea of a paragraph is called the *topic sentence.* This one sentence is the heart of the paragraph, for it either states explicitly or forecasts the main idea. Here are some examples of topic sentences:

The overflow was caused by human error.
Three major factors contributed to the decision to relocate the plant.
The structure of DNA can best be understood by thinking of a spiral staircase.
The Durham Branch is in basic compliance with the new directives.
The computer has led to many changes in aircraft maintenance procedures.

Notice that a topic sentence is merely an assertion. The supporting evidence and elaboration is in the body of the paragraph. In the first example, for instance, readers don't yet know what human error caused the overflow. Presumably, the rest of the paragraph will fill in the details. In the second example, readers are prepared to learn about the three factors that led to the decision to relocate, but they don't yet know what the writer is going to say.

Some writers resist putting a clear topic sentence at the start of the paragraph because they worry that any kind of generalization at the beginning leaves them "exposed." They fear that if they begin the paragraph with "The overflow was caused by human error," they might antagonize those readers who don't want to believe that human error was responsible. For this reason, they think it might be safer to explain the sequence of events and then conclude, "Therefore, the overflow was caused by human error."

I see the situation differently. If the facts are accurate, if the analysis is logical and convincing, the first version will be equally persuasive. And if they're *not,* withholding the conclusion to the end of the paragraph isn't going to help. Writers have to start by believing in what they are saying.

The best place for a topic sentence is the start of the paragraph. Notice, in the following examples, the difference between placing it at the start and placing it at the end. (The topic sentence is underlined.)

In analyzing the yield and viscosity date, we made three assumptions. First, the main objective of the analysis is to maximize yield. Second, the mill permits no more than 1.5 percent rejects. And third, the bleached viscosity should be in the range of 16–20 cps.

The main objective of the analysis is to maximize yield. The mill permits no more than 1.5 percent rejects. The bleached viscosity should be in the range of 16–20 cps. These were our three assumptions in analyzing the yield and viscosity data.

A confession: I stacked the deck a little bit by adding *first, second,* and *third* to clarify the three assumptions. But the point is that this kind of clarification is possible only if the paragraph begins with a topic sentence. In the second version, the three assumptions remain a series of apparently unrelated statements—until the topic sentence gives them a coherent meaning. By that point, some of the readers might have given up in frustration.

Support the Topic Sentence Logically

The purpose of the support—the body of the paragraph—is to make the topic sentence clear and convincing. Sometimes only a few details are necessary.

Each branch of the Society is authorized to hold Planning and Development meetings. To schedule such a meeting, the branch secretary must notify the Society at least 30 days prior to the meeting date. Branch members should receive at least 14 days' notice.

In this paragraph, for example, the writer simply fills in a couple of procedural points.

Sometimes, however, the support carries a heavier load. It has to clarify a difficult idea or defend a controversial one. Because every paragraph is unique, it is impossible to define the exact function of the support. In general, however, it fulfills one of the following five roles:

- to define a key term or idea included in the topic sentence
- to provide examples or illustrations of the situation described in the topic sentence
- to identify factors that led to the situation described in the topic sentence
- to define implications of the situation described in the topic sentence
- to defend the assertion made in the topic sentence

The techniques and patterns used to develop the support are the same used to develop whole documents: chronology, spatial development, classification, partition, general to specific, more important to less important, problem-methods-solution, and cause-effect. (These techniques and patterns are discussed in Chapter 3.)

Emphasize the Coherence of the Paragraph

One task remains after you have written the topic sentence and filled in the support: emphasizing the coherence in the paragraph. Here are three ways to do it:

- Use your key terms—especially the nouns and verbs—consistently. A paragraph about cost overruns probably will use that phrase four or five times. Don't call it *cost overruns* in one place and *financial problems* in another. What seems to you like boring repetition in your writing actually helps your readers follow your train of thought, especially if the subject is very technical.
- Use transitional words and phrases, which point out the direction a thought is following. Their function is similar to that of the topic sentence, for they help your readers concentrate on what you're saying without having to worry about where you're going. Here is a list of the most common logical relationships

between two thoughts and some of the common transitions that express those relationships.

Relationship	*Transitions*
addition	also, and, finally, first (second, etc.), furthermore, in addition, likewise, moreover, similarly
comparison	in the same way, likewise, similarly
contrast	but, however, nevertheless, on the other hand, yet
illustration	for example, for instance, in other words, to illustrate
cause-effect	as a result, because, consequently, hence, so, therefore, thus
time or space	above, around, earlier, later, next, to the right (left, west, etc.), soon, then
summary or conclusion	at last, finally, in conclusion, to conclude, to summarize

In each of the following examples, the first version contains no transitional words or phrases. Notice how much clearer the second version is.

Weak: Computer industry analysts predict double-digit annual sales increases in their industry over the next decade. The Commerce Department predicts 7.6 percent.

Improved: Computer industry analysts predict double-digit annual sales increases in their industry over the next decade. *However,* the Commerce Department predicts 7.6 percent.

Weak: Neurons are not the only kind of cell in the brain. Blood cells supply oxygen and nutrients.

Improved: Nurons are not the only kind of cell in the brain. *For example,* blood cells supply oxygen and nutrients.

In addition to emphasizing the coherence within your paragraphs, make sure the links between them are clear. The best location for the link is the beginning of the new paragraph.

- Use demonstrative pronouns—*this, that, these,* and *those.* In almost all cases, demonstratives should serve as adjectives modifying a noun rather than stand alone as pronouns. In the following examples, notice that a demonstrative pronoun by itself can be confusing.

Unclear: New research techniques are being developed to increase the immune system's strength against secondary infections. These are the subject of a new research effort in California.

What is being studied in California: new research techniques or secondary infections?

Clear: New research techniques are being developed to increase the immune system's strength against secondary infections. *These techniques* are the subject of a new research effort in California.

Even when the context is clear, a demonstrative pronoun used without a noun forces readers back to an earlier idea and therefore interrupts their progress.

Interruptive: Most volume visualization techniques are based on one of five different algorithms. *These* are described in the following paragraphs.

Fluid: Most volume visualization techniques are based on one of five different algorithms. *These techniques* are described in the following paragraphs.

Transitional words and phrases, repetition of key words, and demonstratives cannot *give* your writing coherence; they can only help the reader appreciate the coherence that already exists. Your job is, first, to make sure your writing is coherent and, second, to highlight that coherence.

Keep Paragraphs to a Manageable Length

How long should a paragraph be? As always, the best guide is common sense. A 10-word sentence might be an effective paragraph. Sometimes you will find that you need 250 or 300 words to support your topic sentence adequately. If you think your audience could handle this length, go ahead. But if you think they might have trouble with it, consider breaking the paragraph in half, as the writer has in Figure 6–1:

> The software tools used to support the production of virtual environments fall into two categories: commercial products and research-produced products. The commercial products enable a new researcher to enter the field of virtual environments quickly, but they are quite limited and can quickly become frustrating for the serious researcher. Because commercial products exist primarily to support hardware sold by the software vendor or an allied hardware firm, they offer only a rudimentary environment for creative research in virtual environments.
>
> The research-produced products require more work at the start for the new researcher, but they offer far more opportunities for advanced work. Research-produced software need not support a particular hardware configuration; in fact, it tends to accommodate a broad range of hardware configurations. But although the code can be acquired virtually without cost, the new researcher might have to spend considerable time rewriting it to match hardware needs. Once the interface is smoothed out, however, the researcher can create a virtual environment of unlimited richness.

Figure 6–1 Breaking a Long Paragraph in Half

A strict approach to paragraphing would have required one paragraph, not two, because all the information presented supports the topic sentence that opens the first paragraph. Many readers, in fact, could easily understand a one-paragraph version. However, the writer found a logical place to create a second paragraph and thereby communicated better.

Another writer might have approached the problem differently, making the description of each kind of software a separate paragraph, as in Figure 6–2.

> The software tools used to support the production of virtual environments fall into two categories: commercial products and research-produced products.
>
> The commercial products enable a new researcher to enter the field of virtual environments quickly, but they are quite limited and can quickly become frustrating for the serious researcher. Because commercial products exist primarily to support hardware sold by the software vendor or an allied hardware firm, they offer only a rudimentary environment for creative research in virtual environments.
>
> The research-produced products require more work at the start for the new researcher, but they offer far more opportunities for advanced work. Research-produced software need not support a particular hardware configuration; in fact, it tends to accommodate a broad range of hardware configurations. But although the code can be acquired virtually without cost, the new researcher might have to spend considerable time rewriting it to match hardware needs. Once the interface is smoothed out, however, the researcher can create a virtual environment of unlimited richness.

Figure 6–2 Making Each Point a Separate Paragraph

The original topic sentence becomes a transitional paragraph that leads clearly and logically into the two explanatory paragraphs.

Chapter 7

Writing Better Sentences

This chapter deals with improving the style of your sentences. What is style, and what does it have to do with workplace writing? Style is *how* you say what you say. It is how you sound when someone reads what you have written. Style is important because readers form an impression of you on the basis of how you come across in your writing. You want to appear straightforward, clear, concise, unpretentious, authoritative, and easy to understand—not pompous, unclear, and verbose.

Of course, style alone will not form the impression you make; the information you provide is critical. Yet style is important because if your readers don't like the way you sound, you'll never get an opportunity to impress them with the quality of your thinking.

All this talk about style isn't meant to suggest that you want to sound like someone you aren't; workplace writing is no place for affectation. In fact, the best style is invisible; your readers should not be aware of your presence as a writer. They should not notice that your sentences flow beautifully, even if they do. They should be aware only of the information you are conveying. The reason for this is simple: people don't read workplace writing to appreciate style. They read it because they want to know what you have to say.

Determine the Appropriate Stylistic Guidelines

Before you start to write, it's a good idea to find out if there are any stylistic guidelines that you should follow. This way, you will cut down the time needed for revision. For instance, some companies discourage or forbid the use of the first person—"I" and "we"—in some kinds of documents. An organization's stylistic preferences might be defined explicitly in a company style guide or an outside style manual, such as the *Chicago Manual of Style.* Sometimes, the stylistic preferences are implicit; no style manual exists, but over the years a set of unwritten guidelines has evolved. If this is the case, the best way to learn the house style is to study the documents in the files and ask more-experienced coworkers for advice.

Use the Active and Passive Voices Appropriately

The two voices are active and passive. In an active-voice sentence, the grammatical subject is the person or thing that does the action expressed in the sentence:

Active voice: Smith Construction won the contract for the highway project.

In a passive-voice sentence, the grammatical subject is the recipient of the action expressed in the sentence:

Passive voice: The contract for the highway project was won by Smith Construction.

As you can see, the active voice focuses on the performer of the action, whereas the passive voice focuses on the recipient of the action. Although some books and style programs suggest that the active voice is correct and the passive voice incorrect, it is not a matter of correctness; they just have different functions.

In general, the active voice is preferable. The active voice is always more concise than the passive voice, because the passive requires a compound verb phrase (*was won*) and generally requires a prepositional phrase (*by Smith Construction*). Also, the passive voice can be confusing. If, for instance, you write, "The building was inspected for radon," your readers might be unsure who did it—you or someone else.

But the passive voice is superior to the active voice in four cases:

- when the performer of the action is clearly understood

  Example: Attendees are required to register for the conference by July 15.

  In this sentence from a registration form for a professional conference, it is perfectly clear who is doing the requiring: the conference organizers. It would be unwise to write, "The conference organizers require that attendees register for the conference by July 15," for that would put the emphasis on "conference organizers," rather than on "attendees."

- when the performer of the action is unknown

  Example: The comet was first described in an ancient Egyptian manuscript.

  We don't know who wrote the manuscript.

- when the performer of the action is unimportant

  Example: The materials for the next set of experiments were ordered in March.

It doesn't matter who ordered them.

- when a reference to the performer of the action would be embarrassing, dangerous, or in some other way inappropriate.

Example: Incorrect data were released to the press about the company's toxic emissions.

Your boss did it.

A number of computer programs on style can help you find the passive voice in your writing. With any word-processing program, however, you can search for *is, are, was, were,* and *be,* the forms of the verb *to be* that are most commonly used in passive-voice expressions. In addition, searching for the suffixes *-ed* and *-en* will isolate many of the past participles, which also appear in most passive-voice expressions.

Choose Appropriate Sentence Patterns

Good writers vary their sentence patterns, not only to keep their writing lively, but also to meet the needs of their subject and audience.

There are four basic kinds of sentences:

- simple (one independent clause)

Example: The manager tried to anticipate the problem.

- compound (two independent clauses, linked by a semicolon or by a comma and one of the seven coordinating conjunctions: *and, or, for, nor, so, but,* and *yet*)

Example: The manager tried to anticipate the problem, but he was unsuccessful.

- complex (one independent clause and at least one dependent clause)

  Example: Although the manager tried to anticipate the problem, he was unsuccessful.

- compound-complex (at least two independent clauses and at least one dependent clause)

  Example: Although the manager tried to anticipate the problem, he was unsuccessful, and he decided to halt the project indefinitely.

Two of these four types of sentences are most useful in workplace writing: the simple and the complex. The strength of the simple sentence is that it is clear, direct, and, in general, concise. However, it can communicate only fairly simple ideas. The complex sentence allows for more sophisticated ideas because it creates a single meaning out of two ideas.

The compound sentence, like the complex one, communicates two ideas, but it doesn't combine them into a single idea as effectively. For instance, in the compound sentence "The manager tried to anticipate the problem, but he was unsuccessful," you can see that the writer has created a balance between the two ideas. Sometimes, this can leave the reader confused about which idea is more important, especially when the link is the word *and.* As you are revising, when you see a compound sentence, consider whether you can sharpen it by making it a complex sentence, as in the following example:

Weak: People's taste buds diminish in sensitivity as they age, and processed-food providers monitor demographics carefully and adjust their spice levels accordingly.

Stronger: Because people's taste buds diminish in sensitivity as they age, processed-food providers monitor demographics carefully and adjust their spice levels accordingly.

The revision is stronger because it clarifies the relationship between the sentence's two ideas.

Compound-complex sentences are the most sophisticated kind, but their length and complexity make them inappropriate in many situations.

A basic rule for choosing sentence types is that shorter and less elaborate sentences work best when the subject is complicated or the readers are less knowledgeable about it. When the material is simpler and the readers more comfortable with it, you can accelerate the pace by communicating a greater number of ideas in each sentence.

Focus on the Real Subject

Make sure the subject of the sentence—what you are writing about—is clear and emphatic. Don't hide the subject in a prepositional phrase. Notice in the following examples how prepositional phrases smother the subjects. (The subjects of the sentences are underlined.)

Weak: The purchase of the new robot would improve quality control.
Strong: The new robot would improve quality control.

Weak: The presence of the unidentified gene was detected last week.
Strong: The unidentified gene was detected last week.

A second way to focus on the real subject of the sentence is to cut down on the use of expletives. The constructions—*it is . . . , there is . . . ,* and *there are . . . ,* as well as related forms of the *to be* verb—often can be removed without eliminating any useful information.

Weak: There are many factors that led to the motor damage.
Strong: Many factors led to the motor damage.

Expletives occur naturally in speech, and sometimes they are effective in writing. For instance, it would be hard to find a better way to say "It is raining."

(Notice that my last sentence contains the expletive *it would be.* The alternative, "For instance, finding a better way to say 'It is raining' would be hard," is more difficult to understand because the reader doesn't know where the sentence is going until after it gets there.)

The common culprits here are easy to find with the search function on your computer; most smothered subjects are seen in the vicinity of the preposition *of,* and the expletives can be found by looking for the different forms of the infinitive *to be.*

Focus on the Real Verb

The verb communicates the action in a sentence. Sometimes writers sap the strength of their sentences by turning their verbs into nouns. This process is called *nominalizing,* and the transformed verb is called a *nominalization.* (Did you notice that the word *nominalization* is a nominalization? People who think up these terms have too much time on their hands.) Once the original verb is changed into a noun, the writer has to create a new verb, because sentences need verbs. The new verb is almost always a disappointment. In the following examples, the nominalizations are underlined.

Weak: An <u>analysis</u> of the sample was undertaken.
Strong: The sample was analyzed.

Weak: An <u>investigation</u> of the different options was performed.
Strong: The different options were investigated.

Why do writers nominalize the verbs so often? Most people aren't aware that they are doing it; it's just the way they write when they sit in offices. Nominalizations are just one reflection of the general pomposity of bureaucratic writing. It sounds fancier to say "we performed a damage assessment" than "we assessed the damage." Nominalizations are useful in turning an occasional task into an ongoing project. Instead of just assessing the damage when necessary, you create a damage-assessment schedule, and pretty soon you have a Damage-Assessment Task Force with its own letterhead stationery, secretary, and budget. With any luck, there's a designated parking spot for the chair.

Most nominalizations can be spotted in two ways with the search function. First, they have characteristic suffixes, such as *-tion*, *-ment*, and *-sis*. Second, they are often seen right before *of*.

Use Modifying Elements Effectively

A modifier is a word or phrase that describes some other word in the sentence. For instance, in the phrase "the copier that we bought last year," the words "that we bought last year" modify "the copier"; they tell the reader which copier we're talking about. Workplace writing is full of modifiers, and your job as a writer is to make sure your reader understands whether they are *restrictive* or *nonrestrictive*. In addition, you have to ensure that your reader knows *what* the modifier modifies.

A *restrictive modifier* restricts the meaning of the word or phrase to which it refers. In other words, it identifies it by providing crucial information. In the following examples, the restrictive modifiers are underlined.

The missiles in the museum exhibits are exact replicas of the originals.
Please pay particular attention to the instructions in Part III.

A *nonrestrictive modifier*, on the other hand, just provides extra information about what it refers to. It does not provide crucial, identifying information.

The first mass-produced electric car, the Chevrolet Impact, was released in 1994.
As you leave, stop by the registration area, which is located in the main lobby.

Note that neither kind of modifier requires a pronoun, such as *that* or *which*. However, if you do use a pronoun, use *that* with restrictive modifiers and *which* with nonrestrictive modifiers.

Restrictive: The printer that we bought last week is a Hewlett-Packard.

Nonrestrictive: The printer, which we bought last week, is a Hewlett-Packard.

In this first sentence, the writer is identifying the printer; there must be other printers, at least some of which are not Hewlett-Packards. In the second sentence, the writer is saying two things about the printer: it was purchased last week, and it is a Hewlett-Packard. In other words, the restrictive sentence gives one main piece of information, whereas the nonrestrictive sentence gives two.

Note also that restrictive modifiers are not set off by commas, whereas nonrestrictive modifiers are. This difference suggests one way to tell them apart: say them out loud. If you pause before and after the modifier, it's nonrestrictive; if you don't pause, it's restrictive.

Another way to tell them apart is to cross out the modifier. If the sentence loses its meaning or becomes unclear, it's restrictive. For instance, look at the last set of examples. If you eliminate the restrictive modifier and write "The printer is a Hewlett-Packard," the sentence is unclear because the reader wouldn't know which printer you're referring to. But if you eliminate that phrase from the nonrestrictive sentence, the sentence retains its core meaning, because there is only one printer.

There are two common problems with modifiers: *misplaced modifiers* and *dangling modifiers.*

- A *misplaced modifier* is one that modifies the wrong part of the sentence.

Misplaced: The topic of the meeting is the future of hydroelectric energy in the Red Lion Motel.

Correct: The topic of the meeting in the Red Lion Motel is the future of hydroelectric energy.

In general, keep the modifier near the element it modifies.

- A *dangling modifier* does not refer to anything in the sentence.

Dangling: Analyzing the test report, the data sheet looked incorrect.

The introductory phrase dangles, because the sentence doesn't state who is doing the analyzing. Following are two ways to fix the problem.

Correct: As I was analyzing the test report, the data sheet looked incorrect.
Correct: Analyzing the test report, I thought the data sheet looked incorrect.

Keep Parallel Items Parallel

If you write "We need to order the scanner, purchase the computers, and meet with the network specialist," you have created a parallel list of items. All three tasks are presented the same way: order, purchase, and meet. If you write "We need to order the scanner, purchase the computers, and meeting with the network specialist," the parallelism is violated, because the grammar of the last item in the list does not match that of the first two items.

Parallelism, then, concerns the orderly presentation of logically related units in writing. Why is parallelism important? Sometimes an unparallel presentation can confuse or mislead readers. But mostly it's a matter of sound; you sound more organized and more in control of your information if you present parallel items in a parallel structure.

Parallelism is a general term that refers to many different aspects of writing. This section discusses the most common kinds of parallelism problems.

Lists present a special challenge because you have to line up a string of items. Sometimes it's not easy to make a list of six or eight items line up. Here, for instance, is a typical nonparallel list:

Nonparallel: This is the schedule we hope to follow:

1. writing of preliminary proposal
2. do library research
3. interview with Arway vice president
4. first draft of proposal
5. revision of first draft
6. after we get your approval, publication of final draft

This list is unparallel because the six items are a mixture of noun phrases (items 1, 3, 4, and 5), a verb phrase (item 2), and a noun phrase preceded by a dependent clause (item 6). Following is a parallel version of the same list:

Parallel: This is the schedule we hope to follow:

1. write preliminary proposal
2. do library research
3. interview Arway vice president
4. write first draft of proposal
5. revise first draft of proposal
6. publish final draft, after we get your approval

Don't worry about the grammatical terminology; your ear will tell you when the list is parallel. In this example, I have turned all the items into verb phrases. In general, verb phrases work best for this kind of list because they are concise and uncluttered; it sounds a lot better to write "do library research" than "doing of library research."

Parallelism problems don't appear only in lists; they can plague traditional sentences and paragraphs. Here are some common kinds of parallelism problems:

Unparallel voice: Place the new board in the slot. Then, the board should be pushed in gently until it clicks into place.
Parallel voice: Place the new board in the slot. Then, push the board in gently until it clicks into place.

Unparallel mood: The operator should follow the instructions in Part 2. Do not change the pin settings.
Parallel mood: Follow the instructions in Part 2. Do not change the pin settings.

Unparallel number: The supervisor should be sure they give the technicians plenty of time to ask questions.
Parallel number: Supervisors should be sure they give the technicians plenty of time to ask questions.

Unparallel enumeration: First, be sure to check. . . . Second, align the electrodes. . . . Then, cap the electrodes. . . .
Parallel enumeration: First, be sure to check. . . . Second, align the electrodes. . . . Third, cap the electrodes. . . .

Chapter 8

Choosing the Right Word

This chapter continues the discussion of style by focusing on individual words and phrases. In general, the best advice is to try to be clear and precise and to avoid showing off. Remember, don't worry about individual words and phrases when you are drafting; wait until you revise.

Choose Simple, Clear Words and Phrases

In workplace writing, plain talk is best. If you know what you're talking about, you have nothing to fear. And if you don't know what you're talking about, fancy words and phrases won't fool anyone for more than a few seconds. But it is hard to write simply. It requires constant, sustained concentration, because most of what we hear and read around the office is anything but simple and clear.

How do you know if you've got an overly inflated word? Picture yourself speaking on the phone with your spouse or a good friend.

If you would be embarrassed to have the person hear what you have written, change it.

Pompous: Subsequent to the introduction of the flextime program, a diminution of employee tardiness was manifest.
Plain: Fewer employees come to work late now that we use the flextime program.

Following is a list of some of the most commonly used fancy words and their plain equivalents:

Fancy Word	*Plain Word*
advise	tell
ascertain	learn, find out
attempt (verb)	try
commence	start, begin
demonstrate	show
employ (verb)	use
endeavor (verb)	try
eventuate (verb)	happen
evidence (verb)	show
finalize	end, finish, settle, agree
furnish	provide, give
impact (verb)	affect
initiate	begin
manifest (verb)	show
parameters	variables, conditions
perform	do
prioritize	rank
procure	get, buy
quantify	measure
terminate	end, stop
utilize	use

There is a whole collection of wordy phrases that pop up in workplace writing, too. The same advice applies here: get rid of them.

Fancy Expression	*Plain Expression*
a majority of	most
a number of	some, many
at an early date	soon
at the conclusion of	after, following
at the present time	now
at this point in time	now
based on the fact that	because
despite the fact that	although
due to the fact that	because
during the course of	during
during the time that	during, while
have the capability to	can
in connection with	about, concerning
in light of the fact that	because
in order to	to
in regard to	regarding, about
in the event that	if
in view of the fact that	because
it is often the case that	often
it is our opinion that	we think that
it is our understanding that	we understand that
it is our recommendation that	we recommend that
make reference to	refer to
of the opinion that	think that
on a daily basis	daily
on the grounds that	because
prior to	before
relative to	regarding, about
so as to	to
subsequent to	after
take into consideration	consider
until such time as	until

Watch out for a special category of wordy phrases, the built-in redundancy: *end result*, *any and all*, *each and every*, *completely eliminate*, *very unique*, *major breakthrough*, and *still remain*. Be content to say something once.

Several style programs isolate fancy words and expressions. Of course, with any word-processing program, you can search for those terms that you tend to use inappropriately.

Avoid Unnecessary Jargon

Jargon is shoptalk. To most people, *UPS* stands for United Parcel Service; to an electrical engineer, it's an uninterruptible power source. The word *platform* means one thing to a computer specialist and another to a diver. Although jargon is often ridiculed, it has a useful function: to save time in communicating with other people who understand it. But keep in mind that when you write to someone who doesn't know what the term means—and with e-mails going out to scores of people at once, that is most of the time—jargon can be unclear, intimidating, or even offensive.

Avoid Euphemisms

A euphemism is a polite way of saying something that makes people uncomfortable. Politicians don't raise taxes, they institute revenue enhancements or solicit mandatory contributions. The more uncomfortable the idea, the more euphemisms. In the *Wall Street Journal* (December 7, 1990, p. B1), David Lord, managing editor of *Executive Recruiter News,* presents 48 ways of expressing the idea of firing someone, including *work force imbalance correction, personnel surplus reduction, destaffing, dehiring, decruiting,* and my favorite, *career-change-opportunity creation.* You can't fire me, I initiate employment self-termination.

Avoid Clichés

Good writers avoid clichés. Phrases such as "Go for it" and "I'll be back" are amusing for a while—a very short while. Then they become tiresome. Eventually, they can lose their meaning altogether.

A second problem with clichés is that people frequently get them wrong. For instance, the phrase "I could care less" is often used when the writer means just the opposite.

Following are a cliché-filled sentence and a translation into plain English:

Up to its eyeballs in clichés: Afraid that we were between a rock and a hard place, we decided to throw caution to the wind with a grandstand play that would catch our competition with its pants down.
Clear: Afraid that we were in a difficult situation, we decided on a risky and aggressive move that would surprise our competition.

Avoid Sexist Language

Sexist language favors one sex at the expense of the other. Although sexist language can shortchange males—as in some writing about female-dominated professions such as nursing—in most cases the female is victimized. Common examples include nouns such as *workman* and *chairman* and pronouns as used in the sentence "Each worker is responsible for his work area."

A number of male-gender words have no standard nongender substitutes, and there is simply no graceful way to get around the pronoun *he.* Over the years different organizations have created synthetic pronouns—such as *thon, tey,* and *hir*—but these have never caught on. Some writers use *he/she* or *s/he,* whereas others consider these constructions awkward.

However, many organizations have created guidelines to help their writers reduce the incidence of sexist language. A simple first step is to eliminate the male-gender words. Replace *chairman* with *chair, policemen* with *police officers,* and so forth. Then, reword sentences to eliminate the masculine pronouns:

Sexist: Each worker should make sure he logs in.
Nonsexist: Each worker should make sure to log in.
Nonsexist: Workers should make sure they log in.

If you use a computer, search for *he, man,* and *men,* words and parts of words seen most often in sexist writing. Some style programs search out the most common sexist terms and suggest nonsexist alternatives.

Use Readability Formulas Carefully

A readability formula is a mathematical tool for assessing how difficult a piece of writing will be for a reader. Currently, there are more than one hundred different readability formulas. Many style programs will compute readability formulas for you, or you can calculate them by hand in a few minutes.

In recent years, readability formulas have become quite popular. Some private businesses, for example, stipulate that their public documents achieve a certain readability score; similarly, a number of states require certain scores for leases and other business contracts. The military, too, uses readability formulas.

The only problem is that they are not too helpful. (See Giles [1990] for a review of the literature.) They are based on the idea that long words and sentences are difficult to understand, whereas short words and sentences are easy. This appears to be a logical premise, but it doesn't work for two reasons:

- Word length is measured in syllables, and so *machinery* counts as a difficult word, whereas *quark* counts as an easy one. The same thing applies to sentences; some long ones can be very simple, whereas short ones can be quite complex.
- Readability formulas cannot account for a critical part of the communication: the reader. A word such as *transponder* or *ion* is neither simple nor difficult taken by itself; it is simple for some readers but difficult for others. The same goes for sentences. It all depends on the reader's background and motivation, two factors that a readability formula cannot calculate.

The increased popularity of readability formulas parallels the rise of personal computers. It's easy to have the computer calculate readability, and it's fun. But don't take it too seriously. If the computer tells you that your writing is most appropriate for readers who have had 45 years of schooling, you might want to see if your words and sentences are unnecessarily long. Yet this is a question you ought to ask anyway. Instead of relying on a formula that cannot account for your subject and audience, you would be wise also to try out the document on live people and to assess its readability yourself by reading it out loud over and over again, changing it each time until it sounds easier to understand.

Reference

Giles, T. D. 1990. The readability controversy: A technical writing review. *Journal of Technical Writing and Communication* 20, no. 2: 131–138.

Chapter 9

Creating and Integrating Graphics

The first eight chapters of this book have described the words used in workplace writing. Of course, much written communication in the workplace also involves graphics, such as tables, graphs, diagrams, photographs, maps, and flow charts. This chapter explains some general principles of creating graphics and integrating them into the text. It does not, however, provide detailed information on how to create the many different kinds of graphics; that would require a book by itself. (See Appendix F for a list of good books on graphics.)

Determine Whether You Need Graphics

For some of you, this section heading might look inside out; many technical people, especially scientists and engineers, think visually rather than verbally. They begin to "write" by first creating the graphics; then they write the necessary text to introduce and explain them.

Other technical people start with a verbal plan and then figure out where graphics will help them communicate. Most people do a little of both; they know they want to include some graphics and they know they want to explain some points, so they create the two kinds of communication simultaneously.

It doesn't matter, of course, how you conceive of the document and start to put it together. The important point is to realize that graphics and words complement each other and that you should always be thinking of how you can exploit the special advantages of graphics. Graphics are particularly effective in five ways:

- *Graphics are more interesting to look at than words.* Even a routine table will attract a reader's attention.
- *Graphics are easier to understand and more memorable for many kinds of material.* Try explaining how to assemble a camping tent without using graphics; it's almost impossible.
- *Graphics give you an opportunity to emphasize particular information.* The obvious way to communicate the size difference between Alaska and Rhode Island is to highlight the two states on a standard map.
- *Graphics can save space.* If you need to communicate a lot of information about the two states—their area, percent of land devoted to state and national parks, populations, ethnic mix, and so forth—a verbal description would go on for paragraphs, yet a simple table would do the job efficiently.
- *Graphics are the best way to communicate most kinds of numerical and statistical relationships.* If you want to show, for instance, the relationship between seat-belt use and traffic fatalities over a 10-year period, a paragraph would be almost impossible to follow, but a graph or a table would be clear and meaningful.

As you think about your document or as you review a draft, look for opportunities to include graphics. The following list of words and phrases might help; when you see one of these, ask yourself whether the addition of a graphic would make the text clearer, more interesting, or more memorable:

areas	numbers
categories	parts

causes
changes
compared to
components
consists of
defines
features
fields
functions
hierarchy
mechanism
phases
procedures
processes
relates to
results from
routines
sequences
shape
shares
structure
summary

Determine What Kind of Graphic to Use

Just as an idea can be communicated using different words, it can also be communicated using different kinds of graphics. For instance, the countries of origin of IEEE members could be communicated using a table, a pie chart, a 100 percent bar chart, or a bar graph. Your job is to decide which kind of graphic will work best.

Remember that different kinds of graphics emphasize different points. A table is the least assertive; it presents information without making any comments. A pie chart shows relative proportions, but if there are a lot of slices, the effect is diluted. And if the slices are similar in size, your readers will have a hard time seeing which one is the largest. A bar graph is more like a horse race; the largest quantity appears to be leading the pack. When you think about which kind of graphic to use, don't immediately settle for the first one that comes to mind. Consider which kind will most effectively represent your point. Sometimes the best graphic is an original design, such as the combination of diagrams and a table in Figure 9–1.

And remember that different kinds of readers require different kinds of graphics. Everyone can understand pie charts and standard bar graphs and line graphs, but the average reader would have trouble with a scatter graph or even with a floating bar chart. Make sure you consider the experience and education of your readers.

Don't forget to factor in whether you are speaking or writing to your audience. If you are making an oral presentation, stick to simpler forms. If you are writing a document that your readers can study at their leisure, you can use more sophisticated graphics.

Typical power disturbances

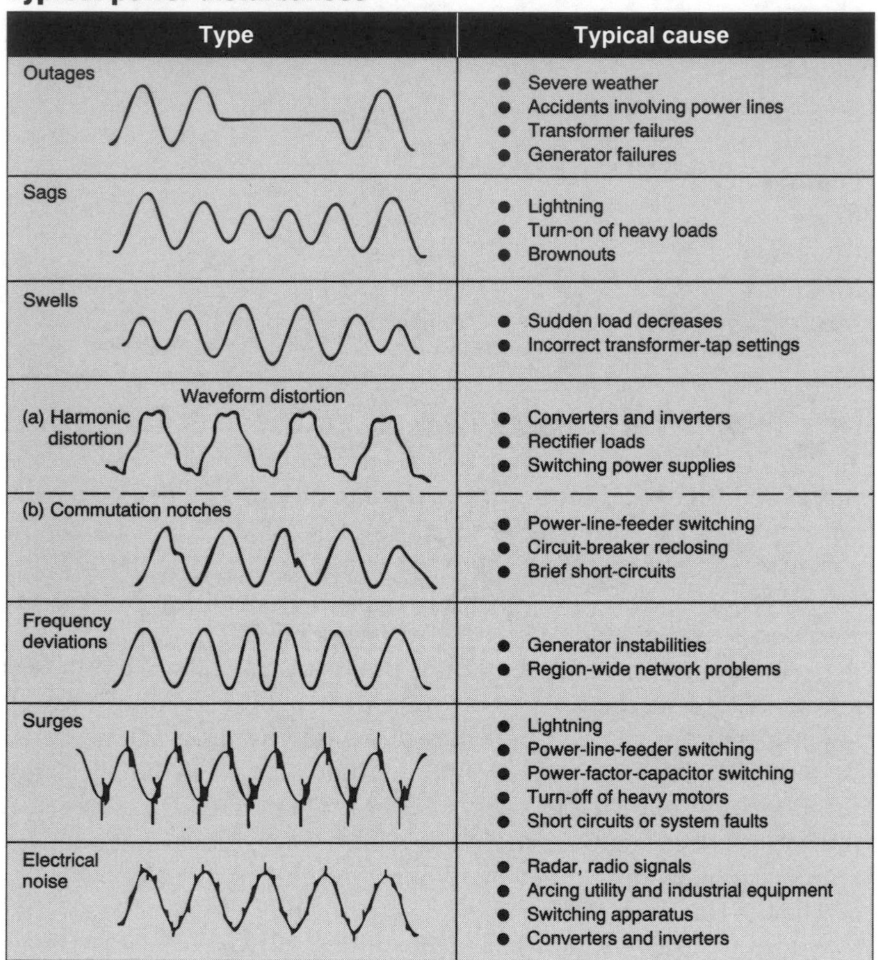

Type	Typical cause
Outages	• Severe weather • Accidents involving power lines • Transformer failures • Generator failures
Sags	• Lightning • Turn-on of heavy loads • Brownouts
Swells	• Sudden load decreases • Incorrect transformer-tap settings
Waveform distortion (a) Harmonic distortion	• Converters and inverters • Rectifier loads • Switching power supplies
(b) Commutation notches	• Power-line-feeder switching • Circuit-breaker reclosing • Brief short-circuits
Frequency deviations	• Generator instabilities • Region-wide network problems
Surges	• Lightning • Power-line-feeder switching • Power-factor-capacitor switching • Turn-off of heavy motors • Short circuits or system faults
Electrical noise	• Radar, radio signals • Arcing utility and industrial equipment • Switching apparatus • Converters and inverters

Figure 9–1 Combined Table and Diagrams (Source: Clemmensen [1993])

Make the Graphic Honest

Writers rarely set out to misrepresent information in graphics, but sometimes it happens because of inexperience. Study and revise your graphics carefully, just as you would your text, to make sure they honestly represent the facts. Two problems are particularly common:

- When you are creating bar graphs and line graphs, try to make

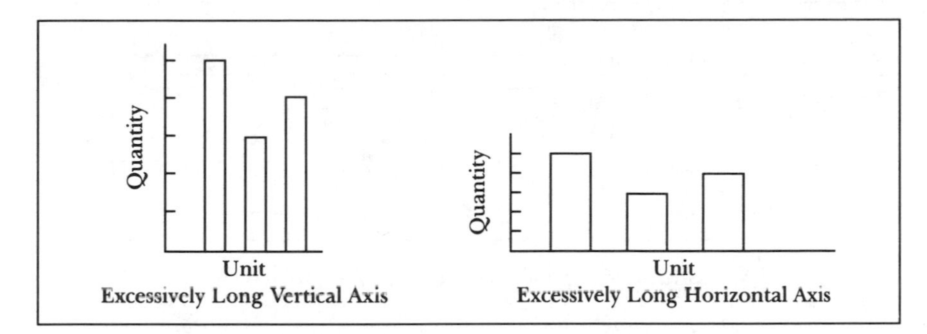

Figure 9–2 Bar Graphs with Excessively Long Axes

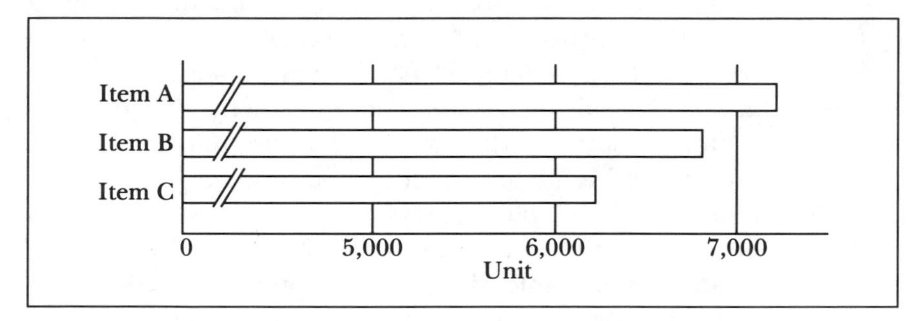

Figure 9–3 A Bar Graph with the Quantity Axis Clearly Broken

the vertical axis approximately two-thirds the length of the horizontal axis. Figure 9–2 shows what happens when one of the axes is disproportionately long. Most graphics software packages use proper proportions, but if yours doesn't, override it.

- If you cannot begin an axis at zero, break it clearly to indicate what you are doing, as shown in Figure 9–3. If you truncate an axis without clearly indicating so, the graph can give a very misleading impression, as shown in Figure 9–4.

Make the Graphic Self-Sufficient

Even though a graphic is part of a larger communication, it has to stand alone because many readers will be looking at it out of context. You don't want your readers to have to flip back and forth between the graphic and your text to understand the graphic. Keep in mind two principles for making the graphic self-sufficient.

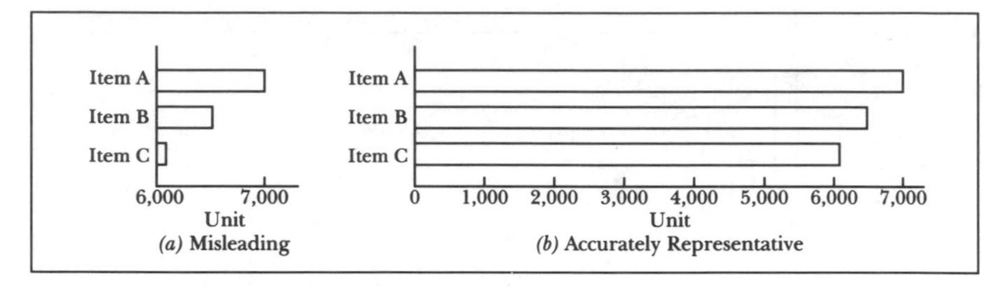

Figure 9–4 Typical Misleading and Accurate Bar Graphs

- *If it is a formal, numbered graphic, give it a clear and informative title.* Include the major factors you are portraying or measuring in the graphics. For instance, "Population Changes in Medford County, 1982–1992" would be better than "Medford County Population." As you think about the title, try turning it into a sentence that begins with the phrase, "This graphic shows. . . ." If the title is sufficiently informative, the sentence will make a clear statement: "This table shows the changes in population in Medford County during the years 1982–1992." By contrast, the sentence "This table shows the population of Medford County" is obviously inadequate, for the title does not indicate the period of time covered by the table. (See Chapter 5 for a discussion of the components of an effective title.)
- *Label the parts of the graphic clearly and honestly.* In a graph, for instance, both the axes should be labeled, with the units indicated clearly. Figure 9–5, a pie chart and segmented bar chart from *IEEE Expert,* shows effective labeling.

In a table, the column headings should be informative. In Table 9–1, notice the use of spanner headings: "Aircraft Total" spans "Units" and "Value"; "Civil" spans two other layers of headings.

One common problem to watch out for in tables: make sure the left-hand column, called the stub, has its own heading. Sometimes writers don't give it a heading at all. Or even worse, they use that space to add a phrase summarizing the headings of all the other columns. In the aircraft table, the stub is accurately labeled "Year."

Finally, if you did not make the graphic itself or if the graphic includes information that you did not create or discover yourself, include a source statement at the bottom of the graphic. This source statement, which is both an ethical obligation and a way to convey

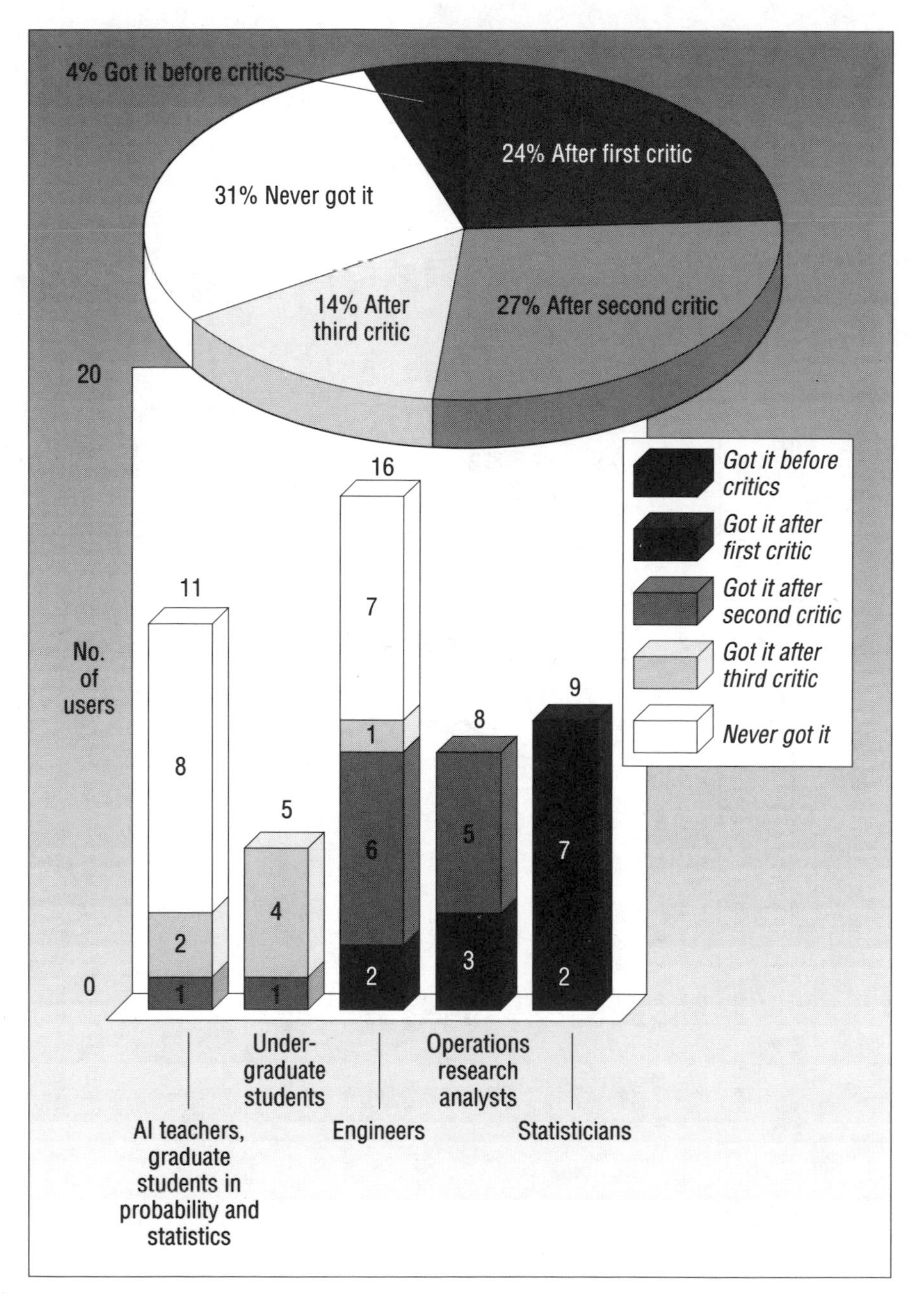

Figure 9–5 Effective Labeling (Source: Silverman [1992])

Table 3: Shipments of Complete U.S. Aircraft, 1971–92

(in millions of dollars except as noted)

Year	Aircraft Total		Civil								Military Total	
			Total		Large Transports		General Aviation[1]		Rotorcraft			
	Units	Value	Units	Value	Units	Value	Units	Value	Units	Value	Units	Value
1971	11,056	6,593	8,142	2,971	223	2,580	7,466	322	453	69	2,914	3,622
1972	13,072	6,220	10,542	3,417	199	2,787	9,774	558	569	72	2,530	2,803
1973	16,509	8,176	14,688	4,814	274	3,873	13,646	828	768	113	1,821	3,362
1974	16,810	8,381	15,297	5,056	322	3,993	14,166	909	809	154	1,513	3,325
1975	17,030	9,136	15,251	5,086	315	3,779	14,072	1,033	864	274	1,779	4,050
1976	17,747	8,888	16,429	4,592	222	3,078	15,450	1,229	757	285	1,318	4,296
1977	19.047	9,031	17,913	4,451	155	2,649	16,910	1,551	848	251	1,134	4,580
1978	19,958	10,177	18,962	6,458	241	4,308	17,817	1,822	904	328	996	3,719
1979	19,297	15,074	18,460	10,644	376	8,030	17,055	2,211	1,029	403	837	4,430
1980	14,681	18,950	13,634	13,058	387	9,895	11,881	2,507	1,366	656	1,047	5,892
1981	11,978	20,093	10,916	13,223	387	9,706	9,457	2,920	1,072	597	1,062	6,870
1982	6,244	19,257	5,085	8,610	232	6,246	4,266	1,999	587	365	1,159	10,647
1983	4,409	22,519	3,356	9,773	262	8,000	2,691	1,470	403	303	1,053	12,746
1984	3,935	21,933	2,999	7,717	185	5,689	2,438	1,698	376	330	936	14,216
1985	3,610	28,386	2,691	10,385	278	8,448	2,029	1,431	384	506	919	18,001
1986	3,258	34,809	2,151	11,857	330	10,308	1,495	1,262	326	287	1,107	22,952
1987	3,010	35,925	1,800	12,148	357	10,507	1,085	1,364	358	277	1,210	23,777
1988	3,254	34,875	1,949	15,855	423	13,603	1,143	1,918	383	334	1,305	19,020
1989	3,675	34,229	2,448	17,129	398	15,074	1,535	1,804	515	251	1,227	17,100
1990	3,418	41,920	2,268	24,476	521	22,215	1,144	2,007	603	254	1,150	17,444
1991[2]	3,244	44,273	2,239	28,226	583	26,200	1,024	1,838	632	188	1,005	16,047
1992[3]	3,132	43,989	2,262	28,926	534	26,829	1,068	1,897	660	200	870	15,063

[1]Excludes off-the-shelf military aircraft.
[2]Estimate.
[3]Forecast.

SOURCE: U.S. Department of Commerce, International Trade Administration (ITA); general aviation (through 1989), General Aviation Manufacturers Association; rotorcraft (through 1989), Aerospace Industries Association. Estimates and forecasts by ITA.

21–6 U.S. Industrial Outlook 1992—Aerospace

Table 9–1 Effective Use of Spanner Headings (Source: *U.S. Industrial Outlook '92: Business Forecasts for 350 Industries,* U.S. Department of Commerce, International Trade Administration, January 1992. Pages 21–6)

your credibility, should be sufficiently detailed to enable your readers to locate the source. For instance, a complete source statement is: "Source: Environmental Protection Agency, *The Impact of Clear Cutting on the Western Cascades,* USEPA Document 8557-498-92 (1992), p. 433."

Determine Where to Put the Graphic

After you have created a graphic, decide where to put it. As a general principle, the more important it is, the closer to the front of the document it should go.

A few graphics are so important that they belong in the summary at the start of the document; you want to be sure the executives and managers see them. (In this case, however, make certain the readers will understand the type of graphic you are presenting.) More commonly, however, graphics are presented in the body or an appendix. When you put a graphic in the body of the document, you are saying that you want your readers to study it because it is necessary for understanding the discussion. When you put a graphic in the appendix, you are saying that you want to make it available for those readers who want to study it, but that you don't feel the information is crucial for an understanding of the discussion.

Too often, writers choose to put graphics in the body when an appendix would be a better choice. If, for example, the graphic is a table giving the full experimental results of the reaction of a biological culture to a range of temperatures, ask yourself whether your readers actually need *all* the data at this point or whether they just need the selected data that you are discussing. If in the body you refer your readers to the full data in the appendix, you keep the pace of the discussion moving along briskly while still enabling those who are interested to find the graphic easily.

Tie the Graphic to the Text

Tying the graphic to the text involves introducing the graphic and explaining its significance.

The best place to introduce the graphic is right before you present it. If you cannot fit the graphic on the same page as the reference to it, make sure the reader can find it easily. If you are certain of the page on which the graphic will appear, include the page number: "The effect of the rainfall on soy yields in Canyon County during the drought in the late 1980s is shown in Table 2, page 8." Often, however, you cannot be certain of where the graphic will appear, either because you don't control the actual production of the document or the document will be revised some time in the future. In these cases, make sure the graphic is numbered (such as Figure 12 or Table 4) and that you have referred to it by its number in the text.

To ensure that your readers understand the meaning of the graphic, ask yourself what its purpose is. Are you trying to provide information that makes or supports a point? If so, explicitly state your point:

> The effect of the rainfall on soy yields in Canyon County during the drought in the late 1980s is shown in Table 2, page 8. Note that the 1987 rainfall, only 32% of normal, was followed in 1988 by a soy yield 55% of normal, whereas the 1989 rainfall, at 78% of normal, was followed in 1990 by a soy yield of 84% of normal. Clearly, there was a linear relationship between rain and soy yields in Ada County during these years; irrigation was of some value, but it could not compensate completely for the drought.

Are you just providing data for purposes of completeness? If your purpose is merely to present data in case your readers are curious, no explanation is necessary:

> The break-even point for the product with a production cost of $45 and a retail price of $100 is 3,000 units. For the break-even points based on other production costs and retail prices, see Appendix 4, page 19.

As a general rule, if the graphic appears in the body of a document, it probably should be explicitly explained; if it appears in an appendix, it probably need not be explicitly explained.

References

Clemmensen, J. M. 1993. Estimating the cost of power quality. *IEEE Spectrum* 30, no. 6: 40–41.

Silverman, B. G. 1992. Building a better critic: Recent empirical results. *IEEE Expert* 7, no. 2: 22.

Chapter 10

Improving Page Design

Writing researchers used to assume that the words and graphics tell the whole story, but now we realize that the appearance of the page—its design—is just as important in communicating meaning to the reader. This chapter discusses page design, starting with two basic aspects—white space and type—and then describing how to design titles, headings, and lists.

As I mentioned in the preface, many large organizations have technical publications departments. When you think about design, consider the professionals in this department, who might be able to work up a few different designs for you or at least help you by offering suggestions or lending you books on the subject.

White space is all the space on the page that doesn't have writing or graphics on it: the margins on all four sides, the indentation, the space between columns on a multicolumn page, and the space between one line and the next. The discussion of white space will cover margins, columns, line spacing, and justification. The discussion of type will cover fonts, families, and sizes.

The advice in this chapter is a mixture of research and common sense. You will find these concepts very easy to apply, and your readers will notice a tremendous improvement in the readability of your documents right away.

Leave Adequate Margins

Margins are the easiest aspect of white space to improve. Writers leave margins on all sides to provide a visual frame for the page. It looks better. For multipage documents, margins are necessary for binding. For standard-size paper, a margin of 1 inch or 1¼ inches is standard. Keep the following points in mind as you design your margins:

- Use smaller margins for smaller pages, such as Quick Reference Guides.
- Use smaller margins when it is absolutely necessary to save space, such as when mailing thousands of copies of a document.
- Use bigger margins when the information is highly technical or when the readers are not very knowledgeable about your subject. Dense pages scare people away.

As I said, this is common sense.

Why do people cram so much writing onto a page? Sometimes they are working on a computer that is set for small margins, and they don't know how to change them or they think they aren't supposed to. And many people use small margins to save paper (and therefore money and trees).

The point about saving trees makes some sense, but a better way to do it is to stop sending photocopies of everything to everybody. The point about saving money is easier to answer: paper is a minuscule portion of the real expense of making a document. A well-designed document is more likely than a poorly designed one to get read in the first place, it reads faster, and it is easier to understand. Do the sensible thing: use wide margins.

Consider a Multicolumn Format

A second aspect of white space to consider is the number of columns to use. For the basic 8½ inch by 11-inch word-processed page, one column is standard, but sometimes you might want to switch to two columns. Keep in mind three factors:

- More information will fit on a multicolumn page, as shown in Figure 10–1.
- The optimum line length for readability is about 60 characters, not the 80 characters of a single column of a standard-width page (Biggs 1968, 40).
- Multicolumn formats give you more flexibility when you use graphics. In a single-column layout, even a narrow graphic takes up a lot of room. In a multicolumn layout, you use only as much space as you need for each graphic; a narrow one can fit in one column, and a wide one can span several columns or the whole page.

For documents such as instructions, a two-column layout lets you put text on the left side and the accompanying graphics on the right, as shown in Figure 10–2.

Figure 10–1 Economical Use of Space in a Multicolumn Layout

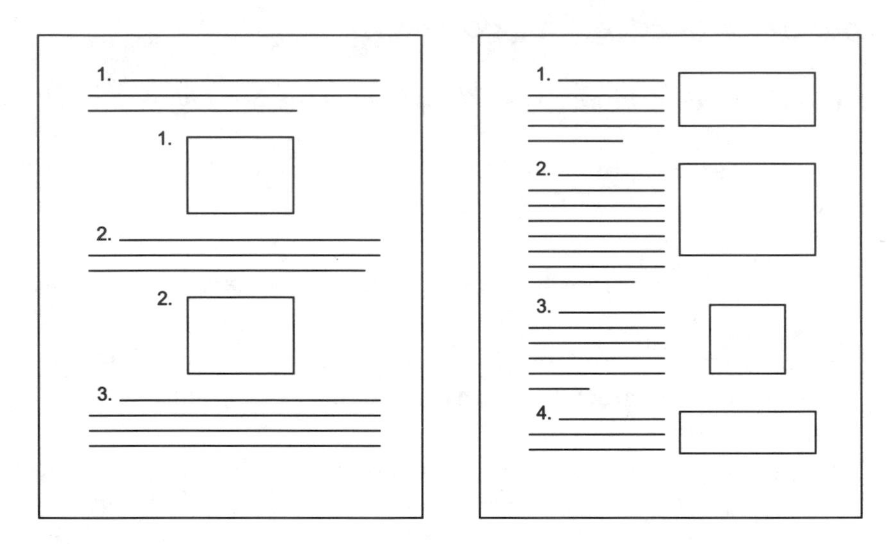

Figure 10–2 Simple Integration of Text and Graphics in a Multicolumn Layout

Use Appropriate Line Spacing

Line spacing, what printers called leading (pronounced "ledding"), refers to the amount of space between lines. How does line spacing affect readability? If it's too small, the page looks dense and intimidating, and the reader's eyes tire quickly. In addition, readers can have trouble locating the next line in the text. If the line spacing is too big, the page looks unfocused, and the reader can become fatigued.

Excessive Leading

Ackley Systems has been contracted by Lojek Electric Cooperative, Inc. (LECI), to design a solid-waste management system for the Bannock County plant, units 3 and 4, to be built in Salem, Idaho. The system will consist of two 1,000-MW pulverized coal-burning

units fitted with high-efficiency electrostatic precipitators and limestone reagent FGD systems. The coal will contain an estimated 3%
sulfur and 10% ash. The station will output approximately 64 TPH
(DWB) of FGD sludge and 24 TPH fly ash at 100% load.

Appropriate Leading

Ackley Systems has been contracted by Lojek Electric Cooperative, Inc. (LECI), to design a solid-waste management system for the Bannock County plant, units 3 and 4, to be built in Salem, Idaho. The system will consist of two 1,000-MW pulverized coal-burning units fitted with high-efficiency electrostatic precipitators and limestone reagent FGD systems. The coal will contain an estimated 3% sulfur and 10% ash. The station will output approximately 64 TPH (DWB) of FGD sludge and 24 TPH fly ash at 100% load.

Inadequate Leading

Ackley Systems has been contracted by Lojek Electric Cooperative, Inc. (LECI), to design a solid-waste management system for the Bannock County plant, units 3 and 4, to be built in Salem, Idaho. The system will consist of two 1,000-MW pulverized coal-burning units fitted with high-efficiency electrostatic precipitators and limestone reagent FGD systems. The coal will contain an estimated 3% sulfur and 10% ash. The station will output approximately 64 TPH (DWB) of FGD sludge and 24 TPH fly ash at 100% load.

Memos and letters are usually single-spaced, with double space between paragraphs. Reports are usually double-spaced, with no extra line spacing between paragraphs. Paragraphs are indented five characters, about a third of an inch. (A better word than *indented* would be *tabbed*, since *indenting* in most word-processing software refers to moving a whole block of text, not just the first line.) For specialized documents such as manuals and some kinds of proposals, line spacing varies according to the difficulty of the subject and the knowledge level of the readers.

Use extra line spacing between sections in a document. For example, if you are typing a single-spaced document with double space between paragraphs, triple-space between sections. In other words, the line spacing *between* sections should be bigger than that *within* sections. You should be able to place your document on the floor, in front of your feet, and see the sections clearly separated from each other. Use extra line spacing, too, to set off graphics or vertical lists.

Notice in Figure 10–3 how the writer has used line spacing to distinguish one section from another. In the left column, the section titled "Power, Distribution, and Specialty Transformers" is set off from the previous section by effective line spacing. The same is true in the right column with "Transformer Orders." Note too that Table 1 is set off from the text with line spacing and that the table that spans the two columns at the bottom of the page is similarly set off with line spacing above it.

Use Appropriate Justification

Justification refers to the alignment of words on the left and right margins of the page. When you use a typewriter, you are writing left justified, ragged right. In other words, the left margin is uniform, but the right margin is not. With a word processor, you can justify both the left and right margins. This is often called full justification.

People frequently use full justification when they get their first computer; it's fun to make it do things a typewriter can't. Some people think full justification looks neat and professional. Is full justification a good idea? If you're using a standard word-processing program, the answer is a clear NO. Full justification is much harder to read than ragged right for three reasons:

- In full-justified text, the computer adds space between words to push them out to the flush right margin. But as readers we are accustomed to uniform spacing. We expect to see a small space separating words and a somewhat bigger space separating sen-

20

Electrical Equipment

The constant-dollar value of shipments by the electrical transmission, distribution, and industrial equipment industries will increase about 1 percent in 1992. During the 1992–96 period, shipments by these industries are expected to rise at a compound annual rate of about 2–2.5 percent.

The four industries covered in this chapter are: power, distribution, and specialty transformers; switchgear; motors and generators; and industrial controls. Transformers are sold primarily to the utility industry; switchgear is used by both utilities and industry; and motors, generators, and industrial controls are for industrial use. Sales of electrical transmission, distribution, and industrial equipment are mainly influenced by levels of industrial production, construction activities, and, to a lesser extent, the demand for electrical appliances.

Before reading this chapter, please see "How to Get the Most Out of This Book" on page 1. It will clarify questions you may have concerning data collection procedures, factors affecting trade data, forecasting methodology, the use of constant dollars, the difference between industry and product data, sources and references, and the Standard Industrial Classification (SIC) system. For topics related to this chapter, see chapters 5 (Construction), 18 (Metalworking Equipment), 19 (Industrial Machinery), and 37 (Household Consumer Durables).

POWER, DISTRIBUTION, AND SPECIALTY TRANSFORMERS

The primary demand for power and distribution transformers is derived from expansion and maintenance in the electric utility industry. According to the Edison Electric Institute, total utility expenditures for transmission and distribution equipment were expected to rise about 5.5 percent in 1991. Although electric utility construction activity increased slightly, the value of transformer industry (SIC 3612) shipments declined about 3 percent in constant dollars.

Electric utility transmission and distribution operating managers agree that parts of the U.S. system are approaching the end of their normal service life of 40 years. Until recently, many utilities have been replacing transmission and distribution equipment installed in the 1930's or earlier.

Transformer Orders

New orders of transformers rated 501 kilovolt-amperes (kVA) and larger totaled 76,322 megavolt-amperes (MVA) in 1990, down from 89,309 MVA in 1989. The order backlog by January 1, 1991 was 44,135 MVA, compared with 71,212 MVA a year earlier. These orders exclude distribution transformers. Transformer units of 35,881 MVA were scheduled for shipment in January 1991 (Table 1).

Table 1: Transformer Shipments, 1989–91
(in megavolt-amperes except as noted)

Power rating in kilovolt-amperes	1989		1990		1991[1]	
	Units	MVA	Units	MVA	Units	MVA
Total[2]	6,207	81,314	5,697	64,833	2,340	35,881
501–10,000	5,121	13,428	4,801	13,161	1,695	5,991
10,001–30,000	672	17,287	581	15,021	438	9,164
30,001–100,000	324	21,320	247	14,606	168	8,724
100,000 and larger	90	29,279	68	22,045	39	12,002

[1]Total scheduled for shipment as of January 1991.
[2]Data for transformers below 500 kVA not collected.
SOURCE: Edison Electric Institute.

Trends and Forecasts: Electrical Equipment (SIC 361, 3621, 3625)
(in millions of dollars except as noted)

Item	1987	1988	1989	1990[1]	1991[1]	1992[2]	Percent Change 1987–88	1988–89	1989–90	1990–91	1991–92
Industry Data											
Value of shipments (1987$)	21,050	22,574	22,590	22,414	22,150	22,326	7.2	0.1	–0.8	–1.2	0.8
3612 Transformers	3,290	3,591	3,568	3,514	3,409	3,341	9.1	–0.6	–1.5	–3.0	–2.0
3613 Switchgear & apparatus	4,907	5,263	4,977	5,030	5,080	5,085	7.3	–5.4	1.1	1.0	0.1
3621 Motors and generators	6,753	7,236	7,254	7,110	6,900	6,940	7.2	0.2	–2.0	–3.0	0.6
3625 Relays and controls	6,101	6,484	6,790	6,760	6,761	6,960	6.3	4.7	–0.4	0.0	2.9

[1]Estimate.
[2]Forecast.
SOURCE: U.S. Department of Commerce: Bureau of the Census; International Trade Administration (ITA). Estimates and forecasts by ITA.

U.S. Industrial Outlook 1992—Electrical Equipment 20–1

Figure 10–3 Effective Use of Line Spacing (Source: *U.S. Industrial Outlook '92: Business Forecasts for 350 Industries*, U.S. Department of Commerce, International Trade Administration, January 1992, Page 95A)

tences. Using full-justified text often results in a bigger space between words than between sentences; the computer, in effect, tricks our eyes. Look, for instance, at the following two passages.

```
This paragraph is typed with full justification. No-
tice that the spacing between  words  is irregular. The
space between   the words "between" and "words" in line
two is bigger than the space between "words" and "is."
```

```
This paragraph, however, is typed with the computer
set to ragged right. You can see that the spaces
between words are the same size.
```

- In full-justified text, the large gaps between words can cause rivers of white space that distract the reader. Look, for example, at the following paragraph:

```
A high-sulfur bituminous  coal gasification plant
is, at this time, more  expensive to  build than
either  a  low-sulfur  bituminous  or  anthracite
plant, but more than   half of its cost  is cleanup
equipment. If this   equipment could be   elimina-
ted, high-sulfur bituminous would be the least ex-
pensive of the three types.
```

Notice how your eyes are drawn to the spaces between words as they flow south from one line to the next.

- In full-justified text, many word-processing programs will break up words at the end of a line. Word breaks slow readers down considerably and increase their fatigue.

Understand the Readability Levels of Type Fonts

The word *font* (often called *typeface*) refers to the design of the letters and punctuation marks. Hundreds of different fonts exist, and for people in the printing business the difference between one font and another is obvious. Most of us, however, would have great difficulty seeing what distinguishes two similar fonts.

You really need to know only a little about fonts. You need to understand that every font makes a different impression on your readers and that some styles of font are easier to read than others.

This paragraph is an example of ITC Zapf Chancery Medium Italic font. You're not likely to see this font in workplace writing; it is used largely for advertisements where the writer wants to evoke an old-world flavor; it recalls the days of hand lettering.

This paragraph is typed in Times Roman, the most popular font. It is very traditional looking, reminiscent of newspaper type from the nineteenth century. Times Roman is an example of what is called a *serif font.* Serifs are the little horizontal and vertical lines at the ends of the letters. For instance, in the *T* in *Times*, the serifs are the little horizontal lines at the bottom and the little vertical lines at the top. The serifs make the text easier to read, because each letter is quite distinct from every other one, and therefore a serif font is best for body text in long documents.

This paragraph is typed in Helvetica, the second-most-popular font. Helvetica is a *sans-serif* font; that is, it doesn't have serifs. Sans-serif fonts have a modern, high-tech look, but because they don't have the extra distinguishing characteristics of serifs, they are somewhat more difficult to read, especially over long stretches of text. Many designers, therefore, choose sans-serif fonts only for titles, headings, and other short passages.

Just as people sometimes use full justification because they have a computer that can do it, people sometimes go overboard with fonts. When they realize they have access to a dozen fonts, they use all of them on every page. This ransom-note effect is usually very unattractive and hard to read. In general, use no more than two fonts in a typical document.

Use the Different Members of a Type Family

A much more effective way to add visual emphasis to your document than using different fonts is to use some of the different members of a font family. Figure 10–4 shows the different members of the Helvetica family of type. When you purchase a font, you get at least several members of the family, usually regular, regular italic, bold, and bold italic. With these variations, you can create different levels of emphasis to represent different hierarchical levels in your document, and the document will have a unified, professional appear-

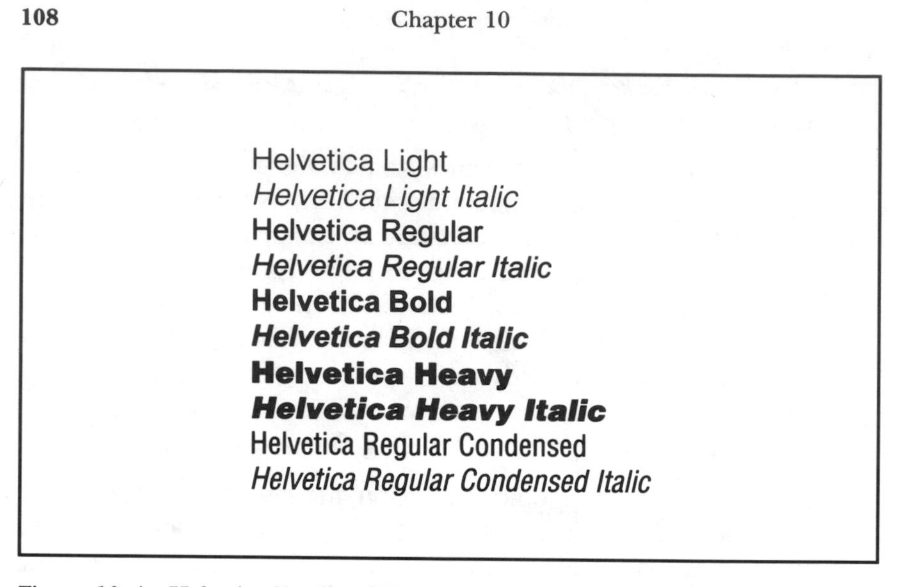

Figure 10–4 Helvetica Family of Type

ance. A common approach is to use bold for a first-level heading, italic for a second-level heading, and regular for a third-level heading. (Avoid underlining to set off headings; recent research by Williams and Spyridakis [1992] shows that underlining does not help headings stand out.)

Use Type Sizes Appropriately

An additional way to add visual emphasis to a document is to vary the type sizes. Size is measured in a unit called a point. Figure 10–5 shows 10 sizes of Times Roman. The standard size for most writing on the job is either 10 point or 12 point.

This paragraph is printed in 10 point. If you use a laser printer or a letter-quality dot-matrix printer, 10 point is legible. If you use a lower-quality dot-matrix, the resolution might not be high enough for easy reading.

Figure 10–5 Ten Sizes of Times Roman

This paragraph is printed in 12 point, which is preferable for lower-quality dot matrix printers. This size is easy to read, but obviously it is less economical than 10 point.

This paragraph is printed in 14 point. This size is appropriate for titles, headings, and special-emphasis text such as warning or danger notes.

Sometimes you will use other sizes. For footnotes, 8 or 9 point is common. For slides and transparencies, 18 or 24 point works well. As is the case with fonts, don't use four or five sizes in a document or else it will end up looking like a car-wash coupon stuck in your windshield wiper at the supermarket.

Use Uppercase and Lowercase

Some writers use uppercase letters exclusively because their equipment is set that way, it saves them time, or they think it adds emphasis to the document.

This practice is a very bad idea, because readers expect an uppercase letter to signify either a proper noun or the beginning of a new sentence. Another problem with using all uppercase letters is that since every uppercase letter is the same height (unlike in lowercase), readers have a harder time distinguishing one letter from another (Poulton 1968). For these two reasons, all-uppercase writing is harder to follow than traditional upper- and lowercase, and reading speed declines considerably.

Design Titles and Headings for Emphasis and Clarity

Chapter 5 explained how titles and headings help your readers understand where you are going. Here the subject is how to design them.

Because the title is the most important single phrase in your document, it should be prominent. For a title page, display the title in a large size, at least 14 point, and center it about one third of the way down the page. For a title at the top of the first page, use a larger size than the most prominent heading in the document. If the most prominent heading is 14 point, 18 point is appropriate. Again, center it.

Headings are designed essentially like titles, except they are not centered but lined up on the left margin or, if there are several levels, tabbed. The general principle is that the more emphatic the level, the closer to the left margin the heading appears. First-level headings generally begin on the left margin, whereas second-level heads are tabbed 5 characters (about one-third of an inch), and third-level heads are tabbed 10 characters (two-thirds of an inch).

It's a good idea to indent the text the same distance as its heading. By doing so, you ensure that the heading remains visually emphatic rather than swallowed up by the text around it.

> **1. First-Level Heading**
> Here is an appropriate design for beginning the first-level text. Notice that it lines up under its heading. Also note that the number 1 hangs out over the left so the reader can see it easily.
>
> 1.1 *Second-Level Heading*
> Notice that the second-level heading and text line up under the first-level text. Also, the second-level text lines up under the second-level heading. This same pattern is used for third and fourth levels.

In designing headings, you might want to vary the line spacing for emphasis.

> *Introduction*
>
> This design emphasizes the heading the most by setting it off with a blank line.

> *Introduction*
> This design is less emphatic.

Introduction. This design is the least emphatic, because the text is run in with the heading.

Design Lists for Clarity

The design of a list is much like that of a heading; the principle is to use indentation to show subordination. With lists, however, there can be complications, because the listed items can be sentence fragments, complete sentences, or a combination of the two.

Style varies widely from one organization to another, but the following guidelines will work well if no other approach is used at your company.

A List Made Up of Fragments

The new warehouse will provide three important benefits:

- better access to the stores
- more modern storage technology
- more modern facilities for the workers

Notice that the listed items begin with lowercase letters and that they are not followed by any punctuation. Some writers would put a period after the word "workers."

A List Made Up of Complete Sentences

The new warehouse will provide three important benefits:

- It will offer better access to the stores.
- It will include more modern storage technology.
- It will offer more modern facilities for the workers.

Because the listed items are complete sentences, they start with uppercase letters and end with periods.

A List Made Up of Items with Both Fragments and Complete Sentences

The new warehousc will provide three important benefits:

- better access to the stores. The average distance is reduced from 2.3 miles to 1.6 miles.
- more modern storage technology. The new warehouse will have an automated command center for more effective and more efficient monitoring and supervision.
- more modern facilities for the workers. The new warehouse will have a modern locker room and lounge facilities.

Each item begins with a fragment that starts with a lowercase letter and ends with a period. Each item also has complete sentences, which begin with uppercase letters and end with periods.

References

Biggs, J. R. 1980. *Basic typography.* New York: Watson-Guptill.

Poulton, E. 1968. Rate of comprehension of an existing teleprinter output and of possible alternatives. *Journal of Applied Psychology*, 52: 16–21.

U.S. Industrial Outlook '92: Business Forecasts for 350 Industries. 1992. U.S. Department of Commerce, International Trade Administration (January).

Williams, T. R. and J. H. Spyridakis. 1992. Visual discriminability of headings in text. *IEEE Transactions on Professional Communication* 35, no. 2: 64–70.

Part II

Applications

Chapter 11

Letters

The business letter is unique among the common kinds of workplace writing in that the form has existed for many centuries, and therefore fairly specific "rules" govern what it looks like. The challenge of writing a good business letter is that although you want it to look conventional, you don't want it to sound old-fashioned. You don't want it to be full of stultifying words such as *herewith* and *aforementioned.* Rather, you want it to sound like a person talking; it should be clear, positive, and professional. This chapter concentrates on the formal requirements of letters as well as strategies for making your letters sound natural.

Learn the Standard Letter Formats

Letter formats have remained almost unchanged for hundreds of years; letters exist from Shakespeare's day that would look perfectly normal today if they were typed instead of written by hand. For some reason, people have resisted various efforts to streamline the format. Therefore, you might as well become familiar with the different formats used in letters; they aren't likely to change any time soon. (One piece of good news courtesy of modern times: with a computer you can easily set up a macro, a template that you can use every time you write a letter.)

Three different formats are used today: full block, modified block, and modified block with indentations. These formats are shown in Figures 11–1, 11–2, and 11–3.

As you can see, the differences among the three formats are minor:

- The *full block format* lines up everything on the same left margin, making it the easiest to type. However, a short letter can look asymmetrical, like an arrowhead traveling east.
- The *modified block format* moves the date line, the complimentary close (the "sincerely" part), and the signature a little to the right of the center of the page. These three elements should line up with each other. Modified block is the most popular of the three formats.
- The *modified block with paragraph indentations format* is just what its name says. The paragraph indentations are purely decorative, for as in all letters the paragraphs, which are single-spaced, are separated by double spaces.

Learn Letter Protocol

Getting the letter format right is simple, especially if you know how to make a macro for it. Only slightly more complicated is the question of letter protocol. By this I mean proper manners. Like most other aspects of manners, letter protocol isn't fully logical.

The first detail you have to think about is how to address the person in the "Dear" section. Is it "Dear Mr. Smith" or "Dear Bob"? Some companies have guidelines that call for using Mr. Smith, regardless of how well you know him. Other companies choose an informal approach; call him Bob, even if you have

May 19, 19xx

Ms. Helen Wright, Director of Operations
Harcourt Dental Group
4667 Woonsocket Drive
Elodie, RI 02095
↕ double space
Dear Ms. Wright:
↕ double space
XXXXXXXXXXXXXXXXXXXXXXXXXXXXXXXXXXX
XXXXXXXXXXXXXXXXXXXX .
↕ double space
XXXXXXXXXXXXXXXXXXXXXXXXXXXXXXXXX
XXXXXXXXXXXXXXXXXXXXXXXXXXXXXXXXXXX
XXXXXXXXXXXXXXXXXXXXXXXXXXXXXXXXXX
XXXXXXXXXXXXXXXXXXXXXXXXXXXXXXXX .
↕ double space
XXXXXXXXXXXXXXXXXXXXXXXXXXXXXXXXXXX
XXXXXXXXXXXXXXXXXXXXXXXXXXXXXXXX
XXXXXXXXXXXXXXX .
↕ double space
Sincerely,

Matthew Ellens
Sales Associate

Enclosure

Figure 11–1 Full Block Format

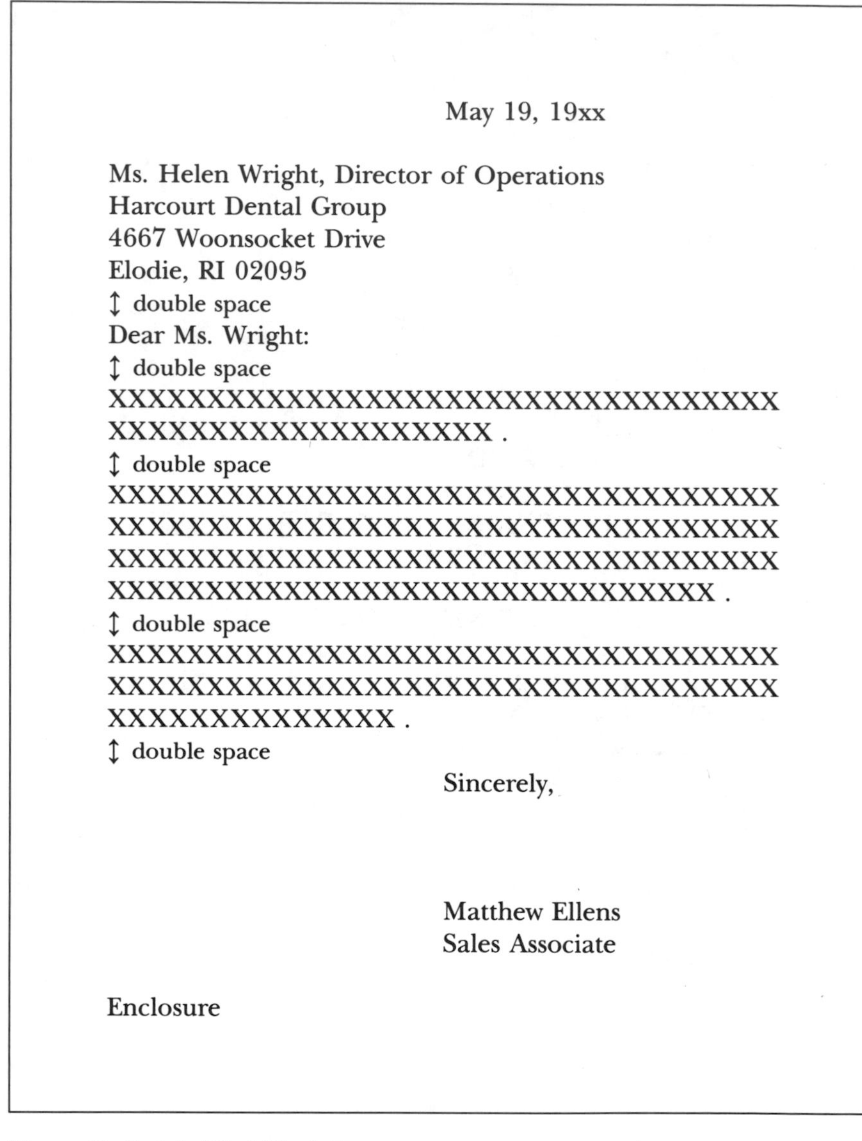

May 19, 19xx

Ms. Helen Wright, Director of Operations
Harcourt Dental Group
4667 Woonsocket Drive
Elodie, RI 02095
↕ double space
Dear Ms. Wright:
↕ double space
XXXXXXXXXXXXXXXXXXXXXXXXXXXXXXXXXXX
XXXXXXXXXXXXXXXXXXXX .
↕ double space
XXXXXXXXXXXXXXXXXXXXXXXXXXXXXXXXXXX
XXXXXXXXXXXXXXXXXXXXXXXXXXXXXXXXXXX
XXXXXXXXXXXXXXXXXXXXXXXXXXXXXXXXXXX
XXXXXXXXXXXXXXXXXXXXXXXXXXXXXXXX .
↕ double space
XXXXXXXXXXXXXXXXXXXXXXXXXXXXXXXXXXX
XXXXXXXXXXXXXXXXXXXXXXXXXXXXXXXXXXX
XXXXXXXXXXXXXXX .
↕ double space

Sincerely,

Matthew Ellens
Sales Associate

Enclosure

Figure 11–2 Modified Block Format

never met him. Still other companies want you to use your judgment. If no stated or unstated guideline applies, use the more formal approach, because it is unlikely to offend your readers,

May 19, 19xx

Ms. Helen Wright, Director of Operations
Harcourt Dental Group
4667 Woonsocket Drive
Elodie, RI 02095
↕ double space
Dear Ms. Wright:
↕ double space
XXXXXXXXXXXXXXXXXXXXXXXXXXXXXX
XXXXXXXXXXXXXXXXXXX .
↕ double space
XXXXXXXXXXXXXXXXXXXXXXXXXXX
XXXXXXXXXXXXXXXXXXXXXXXXXXXXXXX
XXXXXXXXXXXXXXXXXXXXXXXXXXXXXX
XXXXXXXXXXXXXXXXXXXXXXXXXXXXX .
↕ double space
XXXXXXXXXXXXXXXXXXXXXXXXXXXXXX
XXXXXXXXXXXXXXXXXXXXXXXXXXXXXXXXX
XXXXXXXXXXXXXX .
↕ double space

Sincerely,

Matthew Ellens
Sales Associate

Enclosure

Figure 11–3 Modified Block with Paragraph Indentations Format

whereas the chummy sound of "Dear Bob" strikes many people as an inappropriate show of familiarity.

A related detail concerns how to sign off. A few centuries ago, an insufficiently respectful complimentary close could land you in a duel. And as recently as 30 years ago, there were clear differences in familiarity between complimentary closes such as "Sincerely yours" and "Very truly yours." These distinctions have since faded, making most of the forms interchangeable. Appropriate for most business situations are the following six:

Sincerely,
Sincerely yours,
Yours sincerely,
Yours truly,
Yours very truly,
Very truly yours,

When you are writing to high public officials, such as members of congress, you might want to use "Respectfully."

Notice that all the complimentary closes are followed by a comma and that only the first word begins with an uppercase letter.

Forecast the Purpose of the Letter

Like all other kinds of workplace writing, your purpose at the start of a letter is to begin communicating. Specifically, you should link your letter to a previous letter or other kind of communication, and you should clearly announce your purpose.

Help readers understand the occasion for your letter by explicitly referring to their most recent communication with you. This reference not only makes it easier for readers to understand your letter now but also provides a clear record of the communication flow in case someone needs to reconstruct events later.

Also in the first paragraph, announce the purpose of your letter. Use the same kinds of verbs mentioned in Chapter 2, such as "to answer your question about . . . ," or ". . . to let you know the arrangements for. . . ." Don't start to communicate your message yet; save that for the body of the letter. A brief first paragraph is easier to read.

Here are a couple of examples of opening paragraphs that link the letter to a previous communication and announce the purpose:

> Thank you for your letter of March 19. I am glad you had a chance to attend my presentation, and I'm happy to provide more details on our work with manatee protection areas.

> We are pleased to send the information on our SRAM and DRAM products that you requested on July 12.

Notice that these two opening paragraphs are upbeat. Many writers freeze up when they have to write a letter, and the first paragraph comes out, "We have received your request of July 12 for information on our SRAM and DRAM products." The tone here is belligerent, even though writers certainly don't intend it. When inspiration fails you, begin your letter, "Thank you for your . . . " It's hard to go wrong by thanking your reader.

End on a Positive Note

The last paragraph, like the first one, should be upbeat. In the closing paragraph, many writers add personal comments:

> I am looking forward to working with you on the design of the new generator. With the resources of our two companies behind us, I'm sure we'll come up with a first-class design. If you'd like to talk before our meeting on the 14th, please give me a call.

or:

> Again, let me say thanks for the opportunity you afforded me yesterday. I think our system is the best in the industry, and I hope my demonstration made a believer of you. If you have any questions, feel free to contact me.

If you have met your reader's family, you can make your last paragraph a little more personal:

I hope you and your family have a good holiday. Michigan is beautiful this time of year. My preliminary report should be ready by the time you return.

Achieve a Natural Tone

Many writers who would never be rude in a letter have a different problem: they sound like a letter, not like a person.

There is a whole set of words and phrases that over the years have come to be associated with letters. You can live a full lifetime without hearing anyone say "as per your request" in normal conversation, but it seems that half the letters written on company stationery begin that way. When was the last time you felt compelled to say "pursuant to our agreement" to the guy in the next office?

The trick to maintaining a natural tone is to relax. Forget that you're writing a letter. Use the phrases you ordinarily would. Figure 11-4 is a list of letter phrases and plain-talk equivalents.

Letter Phrases	*Plain Talk*
attached please find	attached is
at your earliest convenience	soon
cognizant of	aware that
enclosed please find	enclosed is
endeavor (verb)	try
herewith ("We herewith submit . . .")	______ ("Herewith" doesn't say anything. Skip it.)
in receipt of	we have received
permit me to say	______ (Permission granted. Just say it.)
pursuant to our agreement	as we agreed
wish to advise	______ (The phrase doesn't say anything. Stop wishing. Just say what you want to say.)
the writer ("The writer believes . . .")	"I believe . . ."

Figure 11–4 Letter Phrases and Their Plain-Talk Equivalents

Figure 11–5 consists of two versions of the same letter: one written in letter talk, the other written by a person.

Dear Mr. Smith:

Referring to your letter regarding the problem encountered with your new DataRight personal computer. Our Customer Service Department has just tendered its report to us.

It is their unalterable conclusion that the malfunction was precipitated by the falling of the keyboard onto a hard surface. Reference the marring of the case in the lower right-hand side of the keyboard. We trust you are cognizant of the fact that whereas we guarantee our personal computers for a period of not less than one year against defects in construction and materials, responsibility cannot be assumed by us for a problem such as the one referenced above. We therefore wish to advise, for the reason described hereinabove, that your request for repair without charge cannot be granted.

Permit me to say, however, that the writer would be pleased to see to it that the keyboard is repaired at cost, $25. Your DataRight would then provide you many years of trouble-free service.

Enclosed please find an authorization card. Should we receive this card, we shall endeavor to perform the above-mentioned service and deliver your computer forthwith.

Yours very truly,

Dear Mr. Smith:

Thank you for writing to us about the problem with your new DataRight personal computer.

Our Customer Service Department found a broken connection in the keyboard, which appears to have fallen onto a hard surface, as suggested by the marring of the outer case in the same area, the lower right-hand corner. Although we guarantee our computers for one year against defects in workmanship and materials, we cannot assume responsibility for problems such as this. We cannot, therefore, grant your request to repair the keyboard free of charge.

However, no serious harm was done to the keyboard. We would be happy to fix the broken connection at cost, $25. Your DataRight would then give you many years of trouble-free service.

If you will authorize us to do this work, we will have your computer back to you within four working days. Just fill out the enclosed card and drop it in the mail.

Yours very truly,

Figure 11-5 Two Versions of the Same Letter

Don't be afraid to sound like a person; your readers will appreciate it.

Chapter 12

Memos

The humble memo is the most common type of document written at the workplace; on an average day, you might receive a half dozen and write a few more. Some people don't take much care when they write a memo because it is such a common format and because it is addressed to other people in their own organization.

I think this is a bad idea. Because memos are common and stay in house, you ought to write them carefully. After all, you want people to look forward to reading the memo when they see it's from you; the last thing you want is for them to groan every time they see your name in the "From" section. And if there is any group of readers to whom you should be polite and considerate, it's the other people at your own company; you have to work with them every day.

This chapter describes how to write memos that are easy to read.

Create an Informative Heading

The heading of a memo is the *To-From-Subject-Date* listing at the top. Some companies have preferences about how to indicate the names of the writer and the readers, such as to use the first initial and the last name. Often companies stipulate that the names be alphabetized. It's a good idea to include the job titles with the readers' names; that way, the organizational dynamics of the memo will be clear even after you and some of your readers have moved on to new positions.

When you fill out the subject heading, be sure to include as much information as you can. Start to communicate the point of the memo.

Weak: 200-V Power Supply
Better: 200-V Power Supply Shipping Problems
Better still: Recommended Solutions to the 200-V Power Supply Shipping Problems

As I discussed in Chapter 5, a good title (that's what the subject heading is) communicates the subject and the purpose of the document.

State the Purpose Up Front

As a reader, you know that you want to find out right away why the writer has addressed the memo to you. A succinct statement of purpose at the start of the memo answers the reader's crucial question, "Why are you telling me this?"

Remember that the purpose of a document is not the same as its message. As mentioned in Chapter 2, the purpose of most documents is either to teach the readers something or to affect their attitudes toward something. Here are some examples of purpose statements from memos:

> The purpose of this memo is to inform you of the preliminary test results for the new chip.

This is a request for additional funding to cover the increased costs of the materials for the plant-renovation project.

No writer objects to putting a purpose statement at the front of a memo that simply provides information. Some writers, however, are reluctant to state the purpose up front when they are asking the readers for something, as in the second example. Their thinking is that if you want money, time, or some other resource, it is better to lead up to it by making the case that the request is reasonable and only then state the purpose.

I have asked hundreds of managers and executives for their opinion on this question, and almost all of them say that for in-house writing they would prefer to learn the bad news quickly. It helps them understand where the writer is headed. In addition, this strategy seems more honest; it prevents the impression that the writer is trying to sell readers an idea they don't want to buy.

I also ask managers and executives if they would object if most of the memos they receive began the same way: "The purpose of this memo is to. . . ." Invariably, they say no. They read to find out what the writer has to say, not to be entertained by the writer's creativity. So express the purpose whatever way you want; just make sure you express it.

Include a Summary

For any memo of more than a page, it's a good idea to include a summary right after the purpose statement. Together, these two items enable readers who need only a broad overview to get the information they want—and to get it fast. In addition, the summary provides a useful forecast for those readers who read the whole memo.

In most cases, the summary defines the problem or opportunity that motivated the project, the methods, the important findings, and any conclusions and recommendations. Be sure to present items in the same order in which they will appear in the body of the memo. Figure 12–1 is an example of an effective summary paragraph, preceded by a purpose statement.

Purpose

The purpose of this memo is to clarify the new policy on the use of rebar-eaters and masonry bits.

Summary

Ambiguities in the current policy have come to our attention. This memo defines the old and new policies on the use of rebar-eaters and masonry bits. Of particular importance is a revision of the policy of requisitioning and returning rebar-eaters. Here are the highlights of the changes:

1. Rebar-eaters will be used *only* to cut through rebar.
2. Masonry bits will be used when drilling a hole for an anchor bit.
3. Rebar-eaters will be issued only after a job-description form has been filled out.
4. Rebar-eaters will be returned to the shop immediately after use, but in any case not later than 24 hours after they were issued.

See the discussion below for further details.

Figure 12–1 Effective Summary

The final sentence of this example clarifies that this section is merely a summary of the full discussion that is to follow.

Conclude with an Action Statement

The body of the memo provides the full explanation of the message. You might need an introduction and a background statement before you begin the detailed discussion. Use headings and subheadings liberally throughout the body, and if possible sustain any numbering pattern you established in the summary. For example, if the summary enumerates four major points, use that four-point system in the body.

The last section of many memos is an action statement—a list of future tasks that you or some of your readers or even some other people will carry out. Why set this section off at the end of the memo? Because if you list the tasks throughout the memo, they can easily get lost; the readers might never see them, or they might forget them. Creating a separate section at the end highlights the tasks.

Figure 12-2 is an example of how a writer might conclude a memo:

Future Tasks

By the next department meeting (March 14), I will have:

- telephoned Matt Jenkins . . .
- arranged for the shipment . . .
- coordinated the . . .

Figure 12–2 Effective Action Statement

Figure 12-3 shows how the writer can clearly assign tasks for other people to do.

Action

Here are the jobs I would like each of you to complete by the end of the month:

- Hawkins
    1. Coordinate next year's . . .
    2. Create the . . .
- Bayard
    1. Reassign the four . . .
    2. Troubleshoot the . . .
- Panal
    1. Organize the . . .
    2. Follow up on the . . .
    3. Analyze the . . .

Figure 12–3 Another Effective Action Statement

This list of tasks lacks the graciousness of polite conversation, but it gets the job done.

Figure 12–4, a before-and-after set of memos, exemplifies the points raised in this discussion of memos. The writer is an engineer working in the quality-control section of a computer-chip manufacturing company. She is reporting to her supervisor the results of her analysis of some defective chips.

To: Tom McGregor
From: Sandy Robinson
Subject: Failure Analysis
Date: January 30, 19xx

As per Work Order 422-678, two TQ855-0935 modules were submitted to Quality Control for analysis of failure. I have completed the analysis.

A solderability test was performed on the pads of the printed circuit board of each module. During this test I discovered that each module contains one or two loose pins at the edge connector, with the result being intermittent, open connections.

Usually, loose pins of this kind are caused during the attachment of the connector leads to the printed circuit board. The central problem caused when mounting the leads onto the module is to ensure that the leads are making contact with the printed circuit board edge connector pads on both sides of the module during solder reflow. If the lead and the printed circuit board pad have a gap between them, the solder on the lead will not migrate onto the pad and form a proper solder joint.

The gap between the lead and the printed circuit board is caused by the different thickness of the printed circuit boards and the differences in the lead spacing during the assembly process. In the production of these modules, the leads were pressed onto the module by the operator. This process can spread the spacing between the leads when they are being worked onto the module.

Recently the entire process of attaching the leads was evaluated. Reflow oven profiles were rechecked and new procedures were created to solve this problem. Under this new system, solder is injected onto the printed circuit board, rather than relying exclusively on the solder contained on the leads. This adds thickness to the printed circuit board mounting area, filling in the gap and ensuring a proper solder joint.

For this reason, I don't expect that the problems with the two defective modules will recur. Please call me (x2332) if you have any questions.

To: Tom McGregor, Supervisor, Quality Control
From: Sandy Robinson, Analyst, Quality Control
Subject: Results of Failure Analysis of Two Modules
Date: January 30, 19xx

Purpose

The purpose of this memo is to explain the cause of the failure of two TQ855-0935 modules and to discuss why this kind of failure is unlikely to recur.

Summary

The two modules failed because of loose pins that did not contact cleanly with the printed circuit board connector pads. This problem is unlikely to recur because we have changed the process of applying solder: instead of relying on the solder from the leads, we are now injecting solder onto the printed circuit board, eliminating the gaps around the leads. See the discussion below for details.

Analysis Procedure

As per Work Order 422-678, two TQ855-0935 modules were submitted to Quality Control for analysis of failure.

First I performed a solderability test on the pads of the printed circuit board of each module. During this test I discovered that each module contained one or two loose pins at the edge connector, which would cause intermittent, open connections.

Cause of the Problem with the Two Modules

Usually, loose pins of this kind are caused during the attachment of the connector leads to the printed circuit board. The main goal when mounting the leads onto the module is to ensure that the leads are making contact with the printed circuit board edge connector pads on both sides of the module during solder reflow. If a gap exists between the lead and the printed circuit board pad, the solder on the lead will not migrate onto the pad and form a proper solder joint.

Why do the gaps occur? Usually they are caused by the different thickness of the printed circuit boards and the differences in the lead spacing during the assembly process. In the production of these modules, the leads were pressed onto the module by the operator. This process can spread the spacing between the leads when they are being worked onto the module.

New Procedure That Will Reduce the Gap Problem

Recently the entire process of attaching the leads was evaluated. Reflow oven profiles were rechecked, and new procedures were created to solve this problem. Under this new procedure, solder is injected onto the printed circuit board, rather than relying exclusively on the solder contained on the leads. This new procedure adds thickness to the printed circuit board mounting area, filling in the gap and ensuring a proper solder joint.

For this reason, I don't expect that the problems with the two defective modules will recur. Please call me (x2332) if you have any questions.

Figure 12–4 Two Versions of the Same Memo

The first version of the memo in Figure 12–4 is reasonably clear, but the reader has to read the whole memo before understanding the writer's central points. Without headings to clarify the structure, the reader is forced to rely on the paragraphs themselves, making it harder to follow the memo.

As you can see, it doesn't take much work to make a memo easier to read. Just add headings to clarify the major points; then create a purpose statement and summary for those readers who don't need to read the full memo.

Chapter 13

Minutes

Minutes are the company record of what occurred at a meeting involving company personnel. After minutes are revised and approved—usually at the next meeting—they become legal; that is, they are part of the official documentation of the organization. That is one reason to record them carefully: you don't want to write something unintentionally that will cause you or your company legal problems in the future.

A second reason to be careful is that although people in meetings are not always clear and tactful, the minutes have to be. You want to provide a record that accurately documents the motions and resolutions but still shows everyone in a positive light. People have to work with each other constructively, and a set of minutes can either help smooth over disagreements or perpetuate personal disputes.

How to format the information depends on whether the minutes are informal or formal. Informal minutes are appropriate for meetings attended by a relatively small number of people (fewer than a dozen) from the same department—but definitely from the same organization. Formal minutes are appropriate for meetings attended by a relatively large number of people, some of whom may be from different departments or organizations. Informal minutes are often presented as memos, with the items of business presented as numbered points in the body. Formal minutes are often presented in traditional paragraphs. As usual, you should check to see format preferences at your organization.

Include the Necessary Housekeeping Details

Minutes should include all the housekeeping details—the logistics of the meeting. *Robert's Rules of Order* (Robert and Patnode 1989, 115), the authoritative reference work on parliamentary procedure, recommends including the following information:

- the name of the group or committee that met
- the location, date, and time of the meeting
- the type of meeting (regular or special)
- the presence of the chair and secretary or their substitutes
- the time at which the meeting was adjourned

The *Gregg Reference Manual* (Sabin 1992) recommends also including the names of the people who attended and of those who did not, with a separate list of any guests.

In addition, you should record what action was taken regarding the minutes of the previous meeting (if there was one). For instance, the minutes were read (or distributed) and approved (or amended and approved). You should record any changes to the previous minutes.

Record Events Accurately

The more challenging aspect of writing minutes is recording events accurately. Meetings rarely follow an agenda perfectly, and they can become quite chaotic. In some organizations, minutes are audio taped or even videotaped to help the recording secretary in writing the minutes.

At a minimum, you should record the major topics discussed at the meeting as well as any actions taken. For instance, indicate the names of any reports read or approved, any motions made (and whether they were approved, defeated, or tabled), and any resolutions adopted. When there are votes, record the outcomes. Indicate the names of the persons who made motions, read reports, and so forth; for example, "Bob Yunker presented a report on the July activities of the HVAC department."

Getting the facts straight can sometimes be difficult; if you cannot tape the meeting, you have the right to interrupt the discussion to request a clarification or to ask people to repeat a point. Whenever there is a motion—a proposal to be voted on—you must make sure you have the wording exactly as the proposer stated it. Many recording secretaries routinely ask for the right to read back the resolution before it is voted on.

If, during the meeting, someone presents a document for distribution to the people who are attending the meeting, you can attach it as an appendix without having to rekey it or cut-and-paste it into the minutes.

Record Events Diplomatically

Sometimes it is difficult to separate the substance of the meeting from the emotions. Occasionally, one person will become visibly angry with another, someone will laugh at another person's ideas, or personal comments will be made. What are you to do about these kinds of occurrences? Your job is to try to give some sense of the tone of the meeting but not to record inappropriate or embarrassing behavior; remember, the minutes are written down, and ink is permanent. Therefore, it is appropriate to write:

> After considerable discussion, the motion to add a new position in the Composite Materials Department was defeated by a vote of 7 to 6.

It is inappropriate to write:

> The motion to add a new position in the Composite Materials Department was defeated after a heated argument in which Bill Mitchell complained that management states that it supports his department, but when it comes to action, they're all talk.

Sometimes the best thing a recording secretary can do is choose not to record what was said.

Figure 13–1 is an example of an effective set of minutes.

> Vant Consulting, Inc.
> Monthly Planning Committee Minutes
> Minutes of the February 23, 19xx, Meeting
>
> The meeting was called to order by Chairperson Pines at 3:10 p.m. in the conference room. In attendance were Sipe, Leahy, Robbins, and Zaerr. Evett was absent.
>
> The minutes of the January 21, 19xx, meeting were read. The following correction was made: In paragraph 2 of point 3, "8000 hours" was replaced with "800 hours." The minutes were then unanimously approved.
>
> The subject of the meeting was the planning for the PennDot bid, due May 1, 19xx, to renovate or replace the Woodland Avenue bridge. Chairperson Pines began the meeting by describing the background on the bridge project. The bridge is a thru girder bridge built by the Pennsylvania Railroad Company in 1911. It is currently weight restricted at 2.5 tons, has poor sight distance because of its vertical alignment, and is very narrow. On four occasions in the last two years, the bridge has had to be closed due to hazardous potholes on its wood and bituminous deck surface. Chairperson Pines then asked Sipe to outline the alternative plans that Vant Consulting could propose.

Alternative Plans

Sipe offered four alternatives:

1. Minimal design: Construct a new two-lane thru girder bridge, with no improvement to the existing roadway profile.
2. Minimal improvement: Construct an adjacent box beam bridge with two or four lanes. Slight improvement would be made to the existing roadway profile. Barnum Avenue would be closed by a retaining wall along the western side of Carliner Avenue.
3. Minimal improvement extended: Same as alternative 2, but add roadway improvements beyond Henry Avenue to improve horizontal alignment.
4. Design standard: Same as alternative 3, but improve the roadway to design standards. Barnum Avenue would be closed by a retaining wall along the west side of Carliner Avenue.

Chairperson Pines then asked for comments on the alternatives. Leahy asked whether Sipe had considered if any of the alternatives would be affected by 4(f) regulations regarding the Darby Wildlife Preserve. Sipe responded that he did not know whether the regulations would apply but would check and report to the committee by e-mail. Robbins asked Sipe if there might be a problem with the State Historic Preservation Office because of the age of the bridge. Sipe offered to check on that too.

Chairperson Pines then began a discussion of the environmental factors that would have to be investigated for each of the alternatives.

Environmental Factors

Robbins suggested that soil erosion studies would be necessary. He proposed that existing soil information be collected from the Delaware County Soil Conservation Service and that a modified version of the Universal Soil Loss Equation be used to study the effects of the different alternatives. On the basis of this information, techniques for eliminating or minimizing erosion would be recommended.

Leahy recommended that hydrology studies also be carried out. He suggested that Engineering staff locate the existing stormwater facilities and determine their size, capacity, and condition. Sipe pointed out that Engineering is already overburdened, and with the resignation of Case would not be able to get to it until mid-March. After a lengthy discussion of the implications of waiting for the hydrology studies, Chairperson Pines indicated that he would recommend to the president that the hydrology studies be subcontracted immediately.

Robbins then suggested that a visual effects study of the four alternatives also be added.

There being no more suggestions on environmental studies, Chairperson Pines asked for a vote on the four alternatives and the three environmental studies. They were approved as a package unanimously. Chairman Pines then asked Sipe to flesh out the four alternatives, providing preliminary technical descriptions, cost figures, and timetables. He also asked Zaerr to work with Sipe and provide similar information for the environmental effects. Finally, Chairperson Pines called for a special meeting of the committee on March 12 at 1:30 p.m. to finalize the strategy for the proposal.

New Business

Chairperson Pines asked if there was any new business. There was none.

Chairperson Pines adjourned the meeting at 4:45.

Laura Weddle
Recording Secretary

Figure 13–1 Effective Set of Minutes

Notice that the recording secretary does not end with the phrase "Respectfully submitted," an anachronism that is still seen sometimes. Also notice that while these minutes are relatively formal, the recording secretary uses headings to improve readability.

References

Robert, H. M. and D. Patnode. 1989. *Robert's rules of order.* New York: Berkley.

Sabin, W. 1992. *Gregg reference manual,* 7th ed. Lake Forest, Ill.: Glencoe.

Chapter 14

Procedures and Manuals

This chapter discusses procedures and manuals, two related kinds of documents that explain how to carry out tasks.

A procedure is a brief description of how to carry out a task. For instance, a company might write a procedure for the task of applying for capital expenditures of more than $5,000 or for the task of performing preventive maintenance on a particular piece of equipment. Sometimes the procedure is used as a reference to reinforce oral instructions provided to employees; sometimes the procedure is meant to be self-sufficient. Usually, the procedure covers such topics as the purpose of the task, the schedule for carrying it out, necessary tools and materials, safety instructions, related documentation, steps for carrying out the task, and troubleshooting procedures. Often, these generic topics are addressed in a standard format—safety might be item 2 for every procedure, for instance—so that a reader can pick up any procedure written in the company and know where to turn for a particular topic.

A manual is like a long procedure in that usually it describes how to carry out tasks; common types include installation manuals, training manuals, and maintenance manuals. In addition, there are many different kinds of reference manuals that contain such information as directories of personnel, parts lists, and descriptions of equipment functions. When you buy a commercial software program, you are likely to receive a set of installation instructions, a tutorial manual, and a reference manual. (For information on writing software documentation, see Edmond Weiss' excellent book *How to Write Usable User Documentation* [1991].)

Even though procedures and manuals can vary greatly in size and complexity—a procedure can be less than one page, whereas a manual can be several volumes long—the basic principles are the same.

Before proceeding, it is necessary to add a note about terminology. In some organizations, procedures are called specifications. In other organizations, particularly in engineering disciplines, the word *specification* refers to a proposal submitted to an outside organization, a bid for contract. This use of *specification* is discussed in Chapter 16 on proposals.

Emphasize Safety

The most important thing to remember about instructional writing is the need to emphasize safety. Although some kinds of tasks described in procedures and manuals involve no safety risks, many kinds do, and your chief ethical responsibility is to make sure you have done everything possible to enable your readers to perform the task safely. Emphasizing safety is a three-step process.

Write the Safety Information

In describing how to avoid safety risks, be clear and concise. Don't write complicated sentences, such as "It is required that safety glasses be worn when inside this laboratory." Instead, write simply and directly to the reader: "You must wear safety glasses in this laboratory." Or: "Wear safety glasses in this laboratory." Sometimes a phrase works better than a sentence: "Hard hat area."

Because a typical procedure or manual can contain dozens of comments—both safety comments and nonsafety tips—experts have created different terms to indicate the seriousness of the comment. Unfortunately, terminology is not consistent. For instance, the American National Standards Institute (ANSI) and the U.S. military's MILSPEC publish definitions that differ in some significant ways, and many private companies have their own definitions that don't conform with either ANSI or MILSPEC. The following explanation of four commonly used terms points out the significant differences between ANSI and MILSPEC. The four terms are presented here from most to least serious.

- *Danger.* MILSPEC does not use this term, but for ANSI and many companies *danger* warns the reader of an immediate and serious hazard that will likely be fatal.

  DANGER. EXTREMELY HIGH VOLTAGE. STAND BACK.

- *Warning.* For MILSPEC, *warning* is the most serious level, indicating an action that could result in serious injury or death. For ANSI, it also suggests the potential for serious injury or death. (Among different companies, *warning* ranges from serious injury or death to serious damage to equipment.)

  WARNING: TO PREVENT SERIOUS INJURY TO YOUR ARMS AND HANDS, MAKE SURE THE ARM RESTRAINTS ARE IN PLACE BEFORE YOU OPERATE THIS MACHINE.

- *Caution.* For MILSPEC, *caution* refers to the potential for both equipment damage and long-term health hazards. For ANSI, it indicates the potential for minor or moderate injury. Among companies, *caution* ranges from the potential for moderate injury to serious equipment damage or destruction.

  Caution: Do not use nonrechargeable batteries in this charging unit; they could damage the charging unit.

- *Note.* A note is a tip or suggestion to help the readers carry out the procedure successfully.

  Note: Two different kinds of washers are provided: regular and locking. Be sure to use the locking washers here.

In light of these significant discrepancies, I would recommend the following approach to these four terms if your company or professional organization does not have its own guidelines:

- *Danger.* Likelihood of serious injury, including death.
- *Warning.* Potential for minor, moderate, or serious injury.
- *Caution.* Potential for damage to equipment.
- *Note.* A suggestion or tip on how to carry out a task.

Create a Design for the Safety information

Whether it is in a document or on machinery or equipment, safety information should be designed to be prominent and easy to read. Many organizations use visual symbols to accompany their different levels of comments, but these different symbols are not yet standardized. Determine whether your organization already has some set of symbols that can be used in your document. If you do not have access to any symbols, create different designs for each kind of comment. For instance, warnings could be presented in 18 point type, boldfaced, within a box. Of course, the more critical the safety comment, the larger and more emphatic it should be.

Figure 14–1, clip art from *Arts and Letters,* is typical of symbols used in safety information. The three symbols represent fire danger, electrical danger, and the need to wear safety glasses. Figure 14–2, from an operator's manual for a John Deere lawnmower, shows one company's approach to placing safety warnings on machinery.

Place the Safety Information in Appropriate Locations

What are appropriate locations? This question has no easy answer, of course, because as a writer you cannot control how your audience reads your document. However, a basic principle is to be

Figure 14–1 Symbols Commonly Used in Safety Information (Source: *Arts and Letters Clip Art Handbook*, 15th ed. Dallas: Computer Support Corporation, 1990).

conservative: put in safety information wherever you think the reader is likely to see it, and don't be afraid to repeat yourself. Naturally, you don't want to repeat the same piece of advice in front of each of 20 steps, because that will merely teach your readers to stop paying attention to you. But a reasonable amount of repetition—such as including the same safety comment at the top of each page—may be very effective. If your company's procedure format calls for a safety section near the beginning, put that information there *and* right before the appropriate step in the step-by-step section.

Orient the Reader

More than any other kind of on-the-job writing, procedures and manuals are not read, but are referred to. In other words, you cannot count on your readers' starting at the beginning and reading

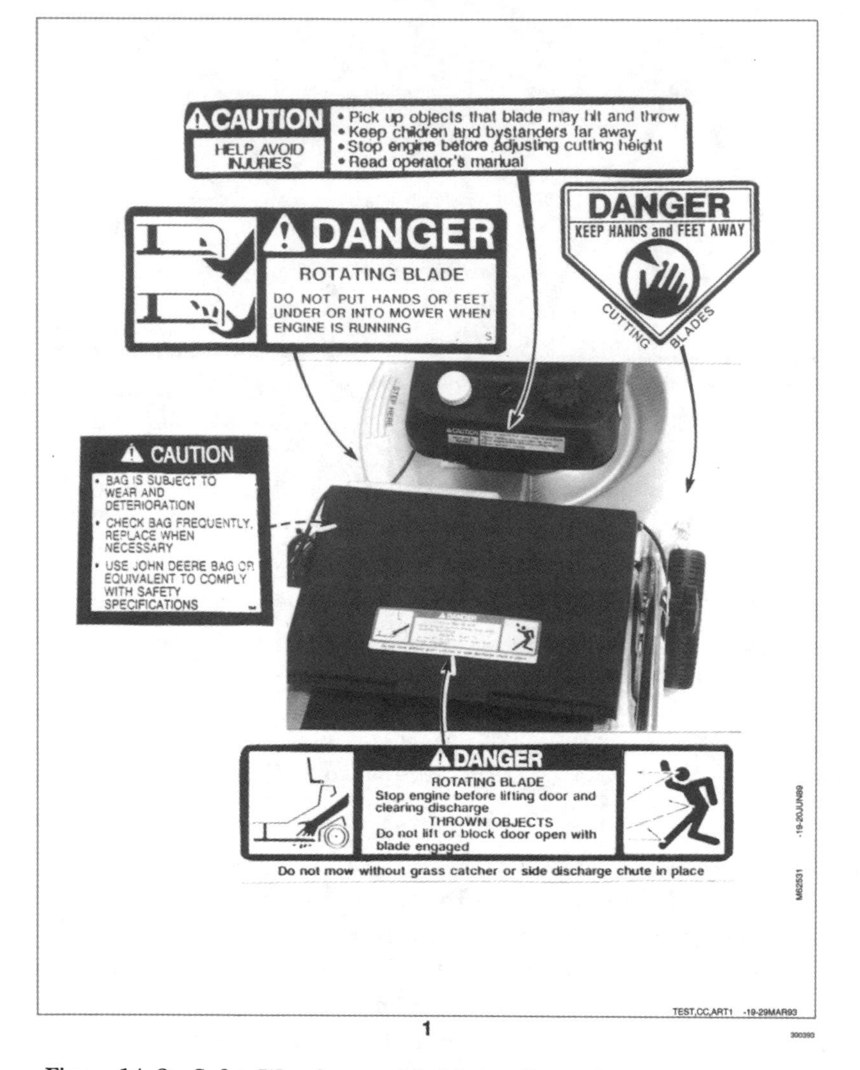

Figure 14–2 Safety Warnings on Machinery (Reproduced by permission of Deere & Company, © 1990. Deere & Company. All rights reserved.)

straight through to the end. (No matter how hard the writers of even a simple set of instructions try, they haven't figured out how to get people to read the whole thing.) For this reason, your job is to help orient your readers so that they know where they are and how to get where they want to be.

The best way to orient readers is to give them an overview at the start of the document and tell them how to navigate. In a brief set of instructions, a paragraph might be all that is necessary:

> Installing your new garage-door opener is a three-step process. First, hang the base unit from the ceiling (section 1). Second, hang the track and attach it to the base unit (section 2). Third, set the codes and check the operation of the opener (section 3). Before you start, make sure you have all the parts of the opener (see page 1), study the safety warnings (page 2), and check that you have the tools you will need (page 2).

For a manual, a how-to-use-this-manual section is common. Your phone book has one of these at the start, as do most software manuals. Figure 14–3, from the introductory chapter of the Microsoft® "Getting Started" manual for its MS-DOS® system, explains how to use the manual. Notice how effectively the informal table directs readers to the discussions they need.

Once your readers are in the manual, you want to help them get from one place to another. Two techniques are used frequently to help with navigation:

- *Cross referencing.* Sometimes readers skip ahead to the discussion of a technical task without having studied the necessary background materials. Cross referencing reminds them of how to get back to the information they need: "For information on how to format different kinds of disks, see page 45."
- *Visual patterns.* It's a good idea to try to create a uniform layout for the document; for example, if the manual is printed like a book, with left-hand and right-hand pages, one pattern might be to have text on the left-hand page and the accompanying graphics on the right. This way, the readers quickly learn to look in the appropriate location for the information they seek. Another technique is to use dividers, tabs, or colored paper to signify different sections of the manual. Again, a phone book is a good example of this technique: we know what the white, yellow, and blue pages mean without having to think about them.

Before you install MS-DOS version 5.0, read this chapter. It includes a list showing you how to find information in this guide and in the *Microsoft MS-DOS User's Guide and Reference*. You'll also find document and keyboard conventions.

Finding the Information You Need

You received two guides with MS-DOS version 5.0: *Microsoft MS-DOS Getting Started* and *Microsoft MS-DOS User's Guide and Reference*. It's best to start with this guide, *Getting Started.*

As you work with MS-DOS version 5.0, the following list can help you quickly locate the information you need.

To learn about	Read
Batch programs	*User's Guide and Reference*, Chapter 10
Command reference	*User's Guide and Reference*, Chapter 14
Computer basics	*User's Guide and Reference*, Chapter 1
Customization of your system	*User's Guide and Reference*, Chapter 11
Device drivers	*User's Guide and Reference*, Chapter 15
Doskey program	*User's Guide and Reference*, Chapter 7
Hard-disk partitions	*User's Guide and Reference*, Chapter 6
Installation of MS-DOS	*Getting Started*, Chapter 2
International support	*User's Guide and Reference*, Chapter 13 and Appendix
Memory optimization	*User's Guide and Reference*, Chapter 12
MS-DOS Editor	*User's Guide and Reference*, Chapter 9
MS-DOS Shell	*User's Guide and Reference*, Chapters 3 and 8
Online Help	*User's Guide and Reference*, Chapters 2 and 3
Troubleshooting	*Getting Started*, Chapter 3
Upper memory area	*User's Guide and Reference*, Chapter 12

Figure 14–3 Excerpt from a How-to-Use-This-Manual Section (Reprinted with permission from Microsoft).

Write Clear Instructions

The heart of most procedures and manuals is the instructional material itself. Keep four points in mind when you write the instructions:

Use the Imperative Mood

This is the technical term for the command style: "Attach the red wire to the pole." The imperative is clearer and shorter than the indicative mood ("You should attach the red wire to the pole," or "The user should attach the red wire to the pole").

Put an Appropriate Amount of Information in Each Step

Don't overload an individual step with too much information. You want your readers to be able to read, understand, and remember the step, so that they can carry out the task successfully and return to the instructions. The following step from a procedure on installing a lock on a door is too long:

> 1. **Marking the Door**
>   Fold the template along the dotted line and tape it to the face and edge of the door about 6 inches above the knob. To drill straight holes, start each one by tapping an awl or finish nail straight in ¼ inch through the marked area on the template. Some templates include marks for doors of different thicknesses; be sure to use the correct mark for the door.

On the other hand, the following step would be too brief:

> 1. **Folding the Template**
>   Fold the template along the dotted line.

Don't Remove the Articles

Although brevity is good, you have to be careful not to take out so much material that the instruction becomes unclear. Sometimes, writers leave out the articles—*a*, *an*, and *the*—because they don't seem to convey any technical information. But they are very important in readability. Look at the following sentence:

> Locate midpoint and draw line.

What does this sentence say? "Locate the midpoint and draw the line" or "Locate the midpoint and the draw line"? In English, a very high percentage of words can functions as nouns, verbs, adjectives, or adverbs. The word *book*, for instance, is a noun, a verb, and an adjective. In most cases, we can figure out which use of the word is intended in a particular sentence. In highly technical writing, however, sometimes we cannot tell without the correct and careful use of an article. Even if the readers can figure out the part of speech without the article, they will get tired reading text without articles:

> Boil water in flask by igniting burner. Resulting steam passes up distillation tube fitted with thermometer and then into water-cooled condenser. Thermometer allows observation of water in flask. When vapor comes in contact with cool walls of inner tube of condenser, it condenses to liquid water.

Don't Remove the Pronouns

Sometimes writers strip out the pronouns to save a few words. Don't. Instead of writing "Check that the bracket is level; then attach with the screws," spell it out more clearly: "Check that the bracket is level; then attach it with the screws." Clearly explaining the step reduces the chance of confusion and makes the writing easier to read.

Add Graphics to Clarify the Text

Because many products are distributed internationally and because literacy is not universal even in developed countries such as the United States, writers of procedures and manuals are relying more and more on graphics. Figure 14–4, from an instruction manual for a cabinet, is typical of modern practice.

Even if you are writing more traditional text-based instructions, try to use graphics to help your readers visualize physical objects such as equipment. And when appropriate, add a graphic to accom-

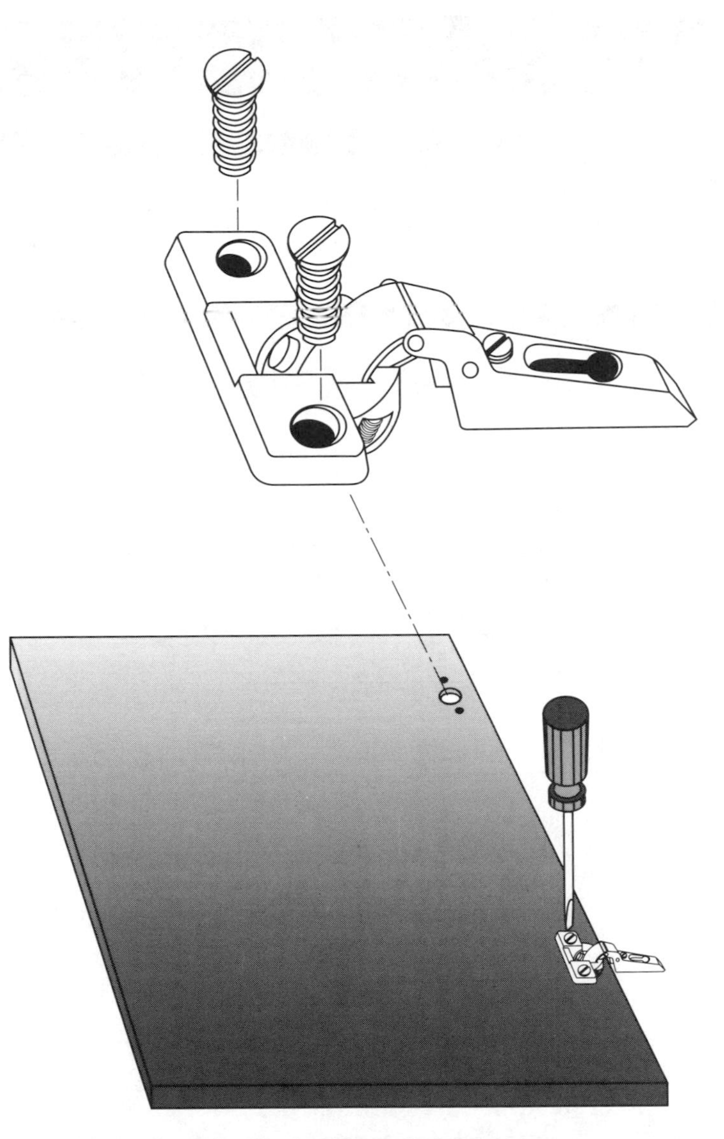

Figure 14–4 Graphics Used in Procedures

pany each step in the process. For instance, a diagram does most of the work in Figure 14–5, an instruction on setting up a CD player; words are used only for conceptual information that would be hard to represent pictorially.

Keep in mind that the best way to show the external surface of an object is not necessarily a photograph. Often, a diagram or

Connections

Notes on connection

- Turn off the amplifier before making connections.
- Be sure to insert the plugs firmly into the jacks. Loose connections may cause hum and noise.
- Leave a little slack in the connecting cord to allow for inadvertent shock or vibration.
- Connect the red plug of the supplied connecting cord to the right-channel jack (R) of the amplifier and the white plug to the left-channel jack (L). Otherwise, the right and left channels will be reversed.

Connection diagram

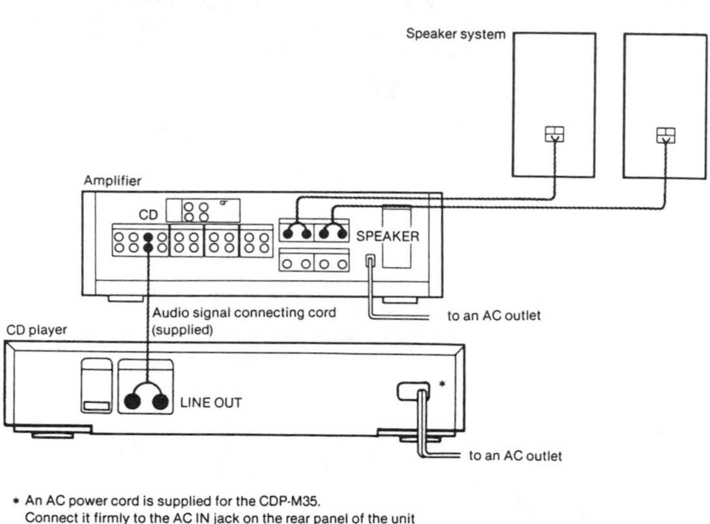

Figure 14–5 Integration of Text and Graphics (Courtesy of Sony Corporation)

sketch will be more effective, because it eliminates the extraneous detail. Consult a book on technical illustration (see the books listed under Graphics in Appendix F) for advice on such kinds of graphics as cutaways and exploded diagrams.

Design the Document for Easy Use in the Field

Because procedures and manuals are likely to be used in the field, they can be exposed to harsh environmental conditions, such as salt water or grease. In addition, the reader might not be able to hold the document with both hands or place it on a table. For these reasons, you should consider five factors about the physical design of the document as you plan it:

- *Size of the document.* Can the document be printed on standard loose-leaf pages, or must it be small enough to fit in a special-size pack or in a pocket?
- *Type of paper.* Some kinds of paper are resistant to water, solvents, chemicals, etc., and therefore might be appropriate for particular kinds of environments.
- *Type of binding.* Binding methods range from the inexpensive and fragile, such as stapling a stack of pages in the upper left-hand corner, to expensive and durable, such as saddle stitching and hardback covers. Some bindings, such as loose-leaf notebooks, make it easy to add and delete pages. Others, such as wire or plastic combs, allow the reader to lay the book flat or to fold pages under to decrease the working space needed.
- *Design of the pages.* If the reader can use a book-format document, in which the document is laid flat, exposing two pages, you have a number of layout options. For instance, the text can be on one page, with the graphics on the facing page, a design used in many formal proposals. However, if the reader will be able to look at only one page at a time, you need a design that integrates the text and graphics on a single page. One such design would call for two columns, with the text in one column and the graphics in the other; another would have a single column, with the graphics placed underneath the corresponding text. (See Chapter 10 on page design for more information.)
- *Size of the type.* Depending on the lighting conditions the reader will be working in, you might have to adjust the size of the type to make the document easy to read.

Check with the technical publications department in your company or an outside print shop (sometimes called a service bureau) for help in assessing these different options and calculating their costs.

Anticipate the Need for Updates

As you plan the procedure or manual, don't forget to think about the need for updates. Few documents will be used indefinitely, without having to be revised periodically. Two aspects of design are particularly important in planning for an update:

- *Binding.* If you anticipate revising a long document, choose a binding such as a loose-leaf binding that enables readers to substitute new pages for old. Otherwise, the whole document will have to be replaced with every update.
- *Pagination.* Documents that will be updated are often paginated by the section or chapter, not consecutively from the first page (like a novel). For instance, the third page in section 7 is page 7-3. This pagination system makes it easier to add a page without having to renumber every page after it. If, for instance, you want to replace page 7-3 with two new pages, you can either number them 7-3 and 7-3a, or you can number them 7-3 and 7-4 and renumber all the remaining pages in Section 7. But you won't have to renumber Section 8.

Include Troubleshooting Tips

Readers hate to be set adrift when things don't work out. An effective conclusion for a procedure or a manual is the troubleshooter's guide, a table that presents typical symptoms, causes, and solutions. Figure 14–6 is an excerpt from the troubleshooter's guide for the Micron 486VL personal computer. Notice that, where appropriate, the Solution column refers the reader to appropriate pages, or to other documents, for more information.

Reference

Weiss, E. 1991. *How to write usable user documentation.* Phoenix, Ariz.: Oryx.

Micron 486VL User's Guide

Common Problems

Introduction . . .

Occasionally while setting up and configuring a computer system, an error might occur or an important element may be forgotten. This section gives information regarding the most frequently encountered problems as well as many quick and simple solutions. If you continue to experience problems and this section fails to alleviate the problem, please feel free to contact Micron technical support (see Appendix A: Contacting Technical Support).

Problems During Boot . . .

Problem	Possible Cause	Solution
When the power switch is turned on, the system does not power up (i.e. the exhaust fan does not run, the power indicator light is not illuminated, the hard disk does not begin to spin, etc.).	The power cable is unplugged or is bad.	Check to ensure the power cable is plugged in correctly and firmly. Try a second power cable, if available, or test the cable for continuity. Replace original cable if it is found to be bad.
	Voltage switch is improperly set.	Check the red voltage switch (located at the rear of the unit near the fan) and ensure that it is set for the proper voltage (115V in the United States and Canada).
	No power at outlet.	Check outlet with another appliance (e.g. a lamp).
	Bad power supply.	Contact Micron technical support.
When the power switch is turned on, the system has power (i.e. the exhaust fan runs, the power indicator light is illuminated, the hard disk begins to spin, etc.), but the system appears dead. Generally the keyboard LEDs will stay on.	The monitor may not be turned on.	Check the monitor to ensure that the power is on. If a problem is suspected with the monitor, please refer to the manual that was provided with it.
	A peripheral card may not be seated correctly on the system board.	Carefully open the system unit (refer to page 2-2, "Opening the Case"). Press down firmly on all peripheral in the expansion slots.
	A peripheral card may be defective.	With the system power shut off, remove one peripheral card, power up the system, and wait a few moments to see if the keyboard light turn off. If they don't, repeat the above steps with the next peripheral card until all have been removed.

4-2

Figure 14–6 Excerpt from a Troubleshooter's Guide (Copyright 1993 Micron Computer, Inc. Used by permission.)

Chapter 15

Formal Elements of Reports

This chapter focuses on the major formal elements found in different kinds of reports. Formal elements, such as tables of contents, executive summaries, and abstracts, are generic units of information that help make your document accessible; an experienced reader knows what kind of information an abstract contains, for example, and therefore turns to it for an answer to a particular kind of question. In creating the formal elements in your reports, therefore, you communicate your information more effectively by shaping it to fit your readers' expectations and then labeling it with terms your readers will immediately understand.

Even though you probably will not include all the formal elements discussed in this chapter in every report you write, it is useful to understand these formal elements so that you can use them whenever you feel they will help you communicate with your readers.

Before you write any of these formal elements, you probably have completed drafting the body of the document. (See Chapter 18 for a discussion of writing the body of a completion report.) The formal elements are discussed here in the most common order of presentation:

transmittal letter
title page
table of contents
abstract
executive summary
[body]
appendices

In writing these elements, however, you are more likely to begin with the body and appendices, delaying the front matter until the end of the drafting stage.

Transmittal Letter

The transmittal letter, which is attached to a report or simply included on top of it, is an important component because it is the first thing the reader sees. Although it generally provides no information that isn't included elsewhere in the report, it is a courteous and graceful way to present the report. Some writers use memos instead of letters when the report is presented in-house, but others prefer the more formal impression made by a letter.

The transmittal letter begins with a brief paragraph defining the subject of the report and indicating that the report is enclosed. The body of the letter is one or two paragraphs communicating the important information: the problem or opportunity and the major findings. The final paragraph is a courteous statement of your willingness to answer any questions raised by the report and perhaps to carry out any other projects for the reader.

A transmittal letter can fulfill a second function: if the report contains an error or omission or something has occurred since the report was assembled, the cover letter is a convenient location to tell your reader about it.

Occasionally, a transmittal letter will also function as the executive summary. A writer who wishes to communicate confidentially to the primary reader, without having any of the other readers see it, will put the sensitive information in the transmittal letter, which is not distributed with the report.

Figure 15–1 is an example of an effective transmittal letter.

March 13, 1993

Captain Lonnie Willis
Engineering Analysis Section
Submarine Antenna Engineering Department
Naval Ship Systems Engineering Station
Philadelphia, PA 19112

Dear Captain Willis:

I am pleased to present the report on nondestructive testing of submarine antenna masts, originally proposed on December 12, 1992. A nondestructive testing system would save money and reduce the chances of mast failure.

To carry out this project I established technical, use and upkeep, and financial criteria, and then tested four leading ultrasound scanners against them. The ultrasonic scanning system from Ultrasonic Testing Inc. met or exceeded all the criteria.

It would cost approximately $40,000 to purchase the Ultrasonic system and train a crew, plus $1,000 per test. It will pay for itself in less than two months.

I strongly recommend that the Submarine Antenna Engineering Department purchase one of the Ultrasonic scanners for a six-month pilot program. If it meets our expectations, I recommend that the scanner be recommended for purchase by all submarine bases.

Please contact me (x3088) if you have any questions; I will be happy to talk with you about them.

Sincerely,

Stephen Moorhatch
Mechanical Engineer II

Enclosure (1)

Figure 15–1 Transmittal Letter (Source: Moorhatch [1993])

Title Page

Although title pages vary from company to company, most have three major elements:

- *Title.* As discussed in Chapter 5, the title should make clear the subject and the purpose of the report, as in "Nondestructive Testing Equipment: A Recommendation for the Failure Analysis Lab." The title should be centered, about one-third of the way down the title page. Often the title appears in large type, such as 18 or 24 point.
- *Names of the writer and the principal reader.* If either of these persons holds a professional title, such as P.E. or Ph.D., include it. The names are commonly written about two-thirds of the way down the page, in smaller type than the title.
- *Date.* The date of submission of the report is added a few lines below the names of the writer and the principal reader.

Table of Contents

Most reports have no index. The table of contents, therefore, is crucial, for it enables your readers to find what they're looking for. A good table of contents should include all the headings and subheadings used in the report. After you put together your table of contents, look at the number of pages between headings; you should have a heading for virtually every page of the report. If you don't, check to see whether you can add additional headings to the report.

One major reason that some tables of contents are insufficiently specific is that they rely too much on generic headings, the general terms that describe an entire class of items. Here is a thoroughly typical "lazy" generic table of contents.

Contents

Introduction	1
Materials	3
Methods	4
Results	7
Recommendations	19
References	22

This contents page isn't much help if a reader is hunting down a specific piece of information.

Once you have made sure your table of contents includes all the headings from the report itself, check to see that you have reproduced the format of the headings accurately. That is, if second-level headings are presented in uppercase letters in the report, they should be presented that way in the table of contents as well; you want to create a visual pattern in the table of contents and sustain it in the report.

Figure 15–2 is an example of an effective table of contents that uses a combination of generic and specific headings as well as standard format options. The report from which it is taken is titled "Nondestructive Testing Methods for Submarine Antenna Masts: A Recommendation."

CONTENTS

Abstract	i
1. **Summary**	1
2. **Introduction**	2
3. **Methods**	4
3.1 *Establish Criteria*	4
3.2 *Select Units to Test*	5
3.2 *Perform Tests*	6
4. **Results**	7
4.1 *Technical Criteria*	7
4.1.1 Detection Abilities	7
4.1.2 Material Thicknesses	8
4.1.3 Mapping Abilities	9
4.2 *Use and Upkeep Criteria*	10
4.2.1 Ease of Use	10
4.2.2 Reliability	11
4.2.3 Maintenance	12
4.3 *Financial Criteria*	14
4.3.1 Initial Cost	14
4.3.2 Operating Costs	15
5. **Conclusion**	16
6. **Recommendation**	17
7. **Appendices**	18
Appendix A: Test Results	18
Appendix B: Initial and Operating Costs	20

Figure 15–2 Table of Contents (Source: Moorhatch [1992])

Some reports require an additional kind of table of contents for the figures, the tables, or for both. This element, called a *list of illustrations*, is a convenience for readers who want to turn directly to a particular graphic. The list of illustrations begins on the same page as the table of contents if there is enough room, or on the next page if there isn't. Generally, figures and tables are listed separately. If the list of illustrations begins on a new page, it is listed in the table of contents.

Figure 15–3 is a typical list of illustrations.

Figures

Figure 1. Mounting Location of Electronic Panel 4
Figure 2. Propulsion Powercircuit Modification 6
Figure 3. Electronic Panel Circuitry 9

Tables

Table 1. Troubleshooting Costs, 1993 4
Table 2. Troubleshooting Costs, 1993-1996 (Projected) 7
Table 3. Repeat No-Power Faults, 1993 15
Table 4. Repeat No-Power Faults, 1993-1996 (Projected) 16

Figure 15–3 List of Illustrations

Abstract

An abstract is a summary addressed to technical readers. It is a guide to the report; the technical person reads it to determine whether or not it is worth the time and effort to read the report. Because it is addressed to technical readers, the abstract can contain technical vocabulary and concepts.

If you are asked to provide an abstract, you should know that there are two basic types: descriptive and informative.

- A *descriptive abstract*, which is sometimes called a *topical, indicative,* or *table-of-contents abstract,* merely lists the topics covered in the report. You can create a descriptive abstract by turning the table of contents into sentences. If a heading in the table of contents reads "Types of Acoustic Monitoring," you can write "The types of acoustic monitoring are discussed." In other words, a descriptive abstract answers the question, "What is the scope of the report?"

- An *informative abstract*, on the other hand, summarizes the important information in the report, emphasizing the results, conclusions, and recommendations. An informative abstract answers the question, "What are the important points made in the report?"

Figure 15–4 is an example of a descriptive abstract based on the report about testing submarine masts mentioned earlier in this chapter.

> A mast failure at sea can cause a potentially catastrophic communication breakdown. Therefore, the Navy has an extensive program to investigate mast damage while the submarine is at port. The present method for determining the extent of damage to a mast involves tapping it to try to hear the waterlogged areas that might indicate a void or crack. However, this method is highly inaccurate, leading to two problems: the scrapping of masts that could be repaired, and the faulty repair of masts that have to be taken out of service again. The research reported here concerns a study to determine whether any commercially available ultrasonic test equipment would improve the accuracy of the mast damage program.

Figure 15–4 Descriptive Abstract (Source: Moorhatch [1993])

Notice that this descriptive abstract sketches in the problem but provides no specific information about the findings of the study.

Figure 15–5 is an informative abstract based on the same report.

In the informative abstract, the writer communicates the major findings of the study. As these examples show, the informative version provides a lot more useful information than the descriptive version does. If you are asked to provide an abstract but are not told which kind, write an informative one. Use the descriptive abstract only when space is at a real premium.

A mast failure at sea can cause a potentially catastrophic communication breakdown. Therefore, the Navy has an extensive program to investigate mast damage while the submarine is at port. The present method for determining the extent of damage to a mast involves tapping it to try to hear the waterlogged areas that might indicate a void or crack. However, this method is highly inaccurate, leading to two problems: the scrapping of masts that could be repaired, and the faulty repair of masts that have to be taken out of service again. The research reported here concerns a study to determine whether any commercially available ultrasonic test equipment would improve the accuracy of the mast damage program. Our conclusion is that the ultrasonic scanner from Ultrasonics Testing Inc. would do an excellent job. This system sends ultrasonic waves from a transducer through the mast to another transducer, determining the number, nature, and locations of any defects. This unit also plots a map of the defects, enabling analysts to determine whether the mast is repairable.

Figure 15–5 Informative Abstract

Executive Summary

Perhaps the single most important component of any kind of report is the executive summary, which is a summary addressed to managers and executives, who presumably are less interested in the technical details of the project and more interested in the managerial aspects. The executive summary and the abstract differ in both purpose and content:

Purpose

An executive summary is meant to be a substitute for the report itself; the manager or executive reads it because he or she doesn't have the time, the expertise, or the need to read the full document. An abstract is meant to be a guide to the report; the technical person reads it to determine whether it is worth the time to read the whole report.

Content

Whereas an abstract summarizes the technical content of the report, an executive summary avoids technical vocabulary and concepts, concentrating instead on managerial concerns:

- What was the problem or opportunity that led to the project? How does this project relate to other current or anticipated projects or initiatives?
- In carrying out the project, did you use any new methods that are themselves of interest? Or did any of the methods involve serious safety or environmental risks?
- How will your findings—the results, conclusions, and recommendations—affect the overall operation of the organization? How much will it cost? What kind of improvements in our operation can we hope to see? When can we hope to see them? Are there any hidden costs, in new hiring, administrative expansion, retraining, down time?

In other words, managers and executives want to know how what you did will affect them in their own areas: hiring or firing of personnel, administrative changes, capital expenditures, regulatory agencies. Think about return on investment and payback periods. Think in terms of money.

The popularity of the executive summary complements the popularity of appendices. Just as the least technical information—the executive summary—is placed in a prominent position, before the body, the most technical information is relegated to the least prominent location, at the end of the document.

The strategy of writing an executive summary is that you are providing the managers and executives with an alternative to the report; the executive summary is their version of the report, and they are actually unlikely to read the rest of the document (for reasons of time, interest, and expertise). Therefore, you have to tell the whole story—from the past through the future—and tell it concisely; many organizations impose a length limitation, such as 300 words or one double-spaced page.

Figure 15–6 is an example of an effective executive summary.

A mast failure at sea can cause a potentially catastrophic communication breakdown. Therefore, the Navy has an extensive program to investigate mast damage while the submarine is at port. The present method for determining the extent of damage to a mast involves tapping it to try to hear the waterlogged areas that might indicate a void or crack. However, this method is highly inaccurate, leading to two problems: the scrapping of masts that could be repaired, and the faulty repair of masts that have to be taken out of service again. In FY 1991–92, the Department of the Navy spent more than $192,000 in unnecessary or faulty repairs of submarine masts; in addition, a mission was delayed two days, at a cost of many thousands of dollars, when a faulty repair job was detected and had to be redone at the last minute.

The research reported here concerns a study to determine whether any commercially available ultrasonic test equipment would improve the accuracy of the mast damage program.

Our conclusion is that the ultrasonic scanner from Ultrasonics Testing Inc. would do an excellent job. This system determines the number, nature, and locations of any defects, enabling analysts to determine whether the mast is repairable. The scanner costs less than $25,000 and is expected to last more than five years. The training costs for personnel amount to less than $16,000, and the cost per analysis of a mast is less than $1,000. The Ultrasonics scanner would pay for itself in less than two months. We recommend that the Ultrasonics scanner be purchased and tested for six months and, if performance meets expectations, the scanner be recommended for purchase by all submarine bases.

Figure 15–6 Executive Summary

Notice that the executive summary focuses less on the technical information—how the scanner works—and more on costs: the cost of the problem and the purchase price and operating costs of the equipment.

Appendices

An appendix is any item attached to the end of a document—a table or figure, a bibliography, a computer printout, supporting letters or other documents, a glossary, or similar item.

Today the average report is shorter than it used to be, but it has far more appendices. What happened? Writers are likely to be much more selective about what will go into the body of the report. Instead of feeling the need to prove every point to their readers in the body, they assert their points and indicate that the full documentation appears in, say, Appendix 3, page 18. In this way, the body of the report is not interrupted by extensive details.

It would be logical to assume that all writers would take advantage of appendices to make the reader's job easier. After all, it's just as easy to put a simple graph in an appendix as it is to include it in the body of the report. But in terms of psychology, the easy way is harder, because it forces the writer to confront the fact that all those hard-earned details are not of primary interest to most of the readers. People don't like to admit things like that.

Try to get into the habit of asking yourself, as you draft the body of the report, whether your readers actually need to read all the information you are including. Is it all necessary if they are to understand you and make their decisions, or is some of it merely extra substantiation, documentation, or amplification?

For example, you are discussing the prices of seven popular laser printers and wish to make the point that Brand C is the least expensive. Why not simply write that Brand C is $175 cheaper than its nearest competitor (and add, if you wish, the price difference between it and the median-priced printer and the most expensive one)? Then, add a parenthetical note cross-referencing the appendix that gives the full listing of the prices. This way, a table doesn't clutter up the body of the report. Remember, when people see a graphic, they stop and study it. Unless they really need to know the prices of all seven models, save them the trouble.

Although information in an appendix is by definition subordinate, appendices are listed in the table of contents.

Reference

Moorhatch, S. 1993. Nondestructive testing methods for submarine antenna masts: A recommendation. Unpublished document.

Chapter 16

Proposals

According to one expert (Holtz 1984), 85 percent of the federal procurement budget, some $200 billion, is spent on goods and services purchased from contractors selected through competitive proposals. And that's just the federal government; hundreds of billions of dollars of goods and services are contracted in private industry. With this much money involved, it's no wonder that proposal writing is the subject of a number of books (see Proposals in Appendix F). For a detailed look at the process of writing proposals and of the different components of proposals, I recommend *How to Create and Present Successful Government Proposals* (Hill and Whalen 1993).

This chapter is a brief overview of the strategy of writing proposals. The two common kinds of reports that often follow the proposal—the progress report and the completion report—are covered in the next two chapters.

A proposal is an offer to perform a task. An *internal proposal*, which you address to people in your own company, is usually an offer to study a problem in the company and report on different options for solving it. When an internal proposal is approved, you receive time and any other necessary resources to carry out the project. An *external proposal*, which you address to readers in a different organization, is also an offer to study a problem and report on different options for solving it, but when an external proposal is awarded, your organization receives a cash payment in exchange for carrying out the task.

This chapter does not focus on the formats, which vary greatly from organization to organization. When you are writing an external proposal, you should consult the organization that issued the request for a proposal (RFP) or the information for bid (IFB) for details on formal requirements. Because a large number of companies might be competing for the award, the issuing organization is likely to have a stringent set of requirements that you must follow carefully. For an internal proposal, formal requirements are usually less stringent, but still it is a good idea to check with more-experienced coworkers to see if any written or unwritten formal guidelines apply.

A note about terminology: in many disciplines, particularly in engineering, the word *specification* is used in place of *proposal*. The strategy for writing specifications and proposals is essentially the same, but specifications are usually written according to guidelines published by state and federal governmental organizations (such as the General Services Administration) and by professional organizations (such as the Associated General Contractors of America). For this reason, specifications are complicated and lengthy documents that are beyond the scope of this book. For detailed discussions, see the books listed under Engineering Specifications in Appendix F.

To further complicate matters, in some fields the word *specification* refers to the RFP or IFB issued by the organization that wants the project done. In such cases, the document that the bidder submits is often called a proposal.

Plan Before You Start Writing

Throughout this book I have talked about the need to plan before you begin writing. In the case of proposal writing, planning is especially important, for two reasons. First, in many cases planning will reveal that you just don't have enough time and money to put together a persuasive document. Second, proposal writing is such a complex enterprise that unless you plan your strategy carefully, you will likely fail, resulting in a lot of lost time and money.

There are three major steps to follow in planning a proposal:

1. *Analyze the readers.* If you have an RFP, study it as if it were a sacred text. What do the issuers say they want, and what can you infer they want? What are their priorities among all the characteristics they list? How will they evaluate the proposal? If you are planning an internal proposal, try to answer the same questions as if you had been furnished an RFP. For more information on studying RFPs, see the next section.
2. *Analyze your resources.* Once you have decided what your readers really want, determine whether your organization can fulfill those needs. If they want a low-cost solution but your strength is high quality, you might wisely decide at this point not to pursue the contract at all. If you think their needs match your strengths, start to plan in detail: what do they need delivered, and when? What resources—personnel, equipment, management plans, etc.—will you need to acquire or create? How much time (and therefore money) will you need to put together a good proposal?
3. *Create a plan for writing the proposal.* If you have concluded that the proposal would have a good chance of success, write a detailed plan for creating the proposal. Include people, tasks, objects, and time; that is, you should specify who is to do what task, what the different people will need to carry out their tasks, and when the tasks must be completed. Naturally, the more detailed this plan, the less likely that you will overlook something important.

With this planning complete, you can start to write the proposal itself. The following sections describe a strategy for blocking out the proposal.

Show that You Understand the Readers' Needs

The first and most important challenge in writing any internal or external proposal is to show that you understand the readers' needs. This might seem like an obvious point, but people who review proposals say that the biggest weakness they see is that writers don't address what the offering organization says is the problem. Instead, they gloss over the needs and concentrate on the details of what they want to do. Often, the plan is detailed and logical, but it doesn't respond to the problem.

Why do writers fail so fundamentally? Three explanations seem to be at work:

- *The writer thinks the readers already understand the needs, so there is no need to waste time repeating them.* The readers *might* understand the needs, but that's not the point. The reason to explain the readers' needs is not to tell them something they don't already know (although sometimes readers of an internal proposal aren't even aware that a problem exists). Rather, the reason to explain the needs is to make clear that *the writer* understands them. Otherwise, no matter how impressive the rest of the proposal, readers will wonder if the completed project will ever solve the problem.
- *The writer can't figure out the reader's needs.* Some RFPs and IFBs are unclear, either because they are poorly written, because the issuing organization doesn't understand its own needs, or both. In such cases, the writer tries to figure out the readers' needs but simply fails.
- *The writer really doesn't know how to respond to the readers' needs.* Sometimes the writer realizes that his or her proposal won't meet the readers' needs but hopes to come closer than anyone else's. Sometimes, of course, the writer is thoroughly cynical, merely hoping to win the contract for some project even though it has no chance of solving the readers' problems.

Your readers, therefore, will be looking for a specific answer to the question, How will you help me solve my problem? If you're not sure you can provide a specific answer, go back to the RFP or your analysis of the reader's needs.

Notice in Figure 16–1 how effectively the writer shows an understanding of the reader's needs. The writer, an engineer working for a regional rail system, is writing an internal proposal for a project to reduce repair costs for the rail cars.

At the April meeting of the Shop Priorities Committee, it was concluded that since the problem with the Fault Lights in the General Electric cars is putting a severe strain on shop personnel and resources, a high priority should be given to determining the cause of the problem and exploring ways to solve it. The following discussion outlines our understanding of the problem at this point.

In a train consisting of General Electric cars, the first indication of trouble is the lighting of the Fault Light in the engineer's cab of the head car. If the General Electric car is in a draft of several cars, the other cars are now towing the dead car, which is an inefficient use of power and places a greater load on the live cars, reducing the life of the traction motors and power transformers.

Although several different problems can trip the Fault Light (such as the tripping of a ground relay or a hot transformer relay), only the general No Power problem is resetting; that is, once the Fault Light trips, all local fault indicators return to operational state, making it impossible to track the exact location of the problem easily.

When the Fault Light trips, the engineer is required to notify the Mechanical Desk operator by radio. The engineer indicates the number of cars in the train, which car is not taking power, and the maximum speed of the train. The Mechanical Desk operator uses this information to determine what course of action to take. If the train is able to maintain its schedule, the operator tells the engineer to keep the train in service and informs the appropriate yard personnel that a train with a No Power fault will be arriving and will need to be repaired.

When the train arrives in the yard, two mechanics are assigned to check it for problems. They check all the propulsion circuits, but in over 90 percent of the cases they find no problem with the car. They reset the No Power fault relay and return the car to service. This process takes the mechanics approximately four hours. In the past month alone, this process has been carried out ten times, at a cost of over $1,000 in shop charges alone.

Figure 16–1 Showing an Understanding of the Readers' Needs (Source: Schiller [1992])

Notice the clarity and precision of this description; it shows that the writer has a good understanding of the problem.

Propose a Clear, Specific Technical Plan for the Project

After you have made the case that you understand your readers' needs, the rest of the proposal must provide a clear, specific plan of what you would do if the project were approved.

To understand this point, simply think of what you want when you solicit bids for some work on your house. You want the contractors to tell you exactly what they will do, why they want to do it that way, how they will do it, where they will do the work, and when they will do it. In addition, you want to know more about the people who will be doing the work. These journalistic questions—the *who, what, when, where, why,* and *how*—help you understand the kind of information you have to provide in the proposal.

In general, these journalistic questions concern two aspects of your proposal: the technical plan and your own professionalism and credibility. This section covers the technical plan; the next section will cover professionalism and credibility.

What

What precisely are you going to provide? If it is a report, describe it in detail. What will be its scope? Purpose? Organization? Length? (See the discussion of introductions in Chapter 5 for a description of the kinds of questions you should address.) If it is a physical product, describe it fully, providing all the physical and operating specifications.

Why

This is a critical question: why do you want to do the project this way? If there were only one way to do it, you would have nothing to say, but almost always several ways exist, so you must justify your strategy for solving the problem. Often, you will refer to the professional literature, the trade magazines, and your own experience in similar projects to make the case that your way to approach the project is most likely to succeed.

How

If the *why* question calls for strategy, the *how* question calls for tactics: what will be your methods for carrying out the project? The more detailed you can be in describing your methods, the more professional you appear; you seem confident, willing to let the reader examine your ideas.

Figure 16–2, from the proposal by the engineer for the regional rail system, shows a clear plan of action.

> I propose building and installing an electronic panel to indicate the state of pertinent relays at the time of the No Power fault. This panel would hold information in memory to be used to determine the actual problem that occurred on the car. The panel could be built at the Powelton Avenue Yard by yard mechanics and installed during the car's layover. The panel could be monitored by the Mechanical Department when a failed car comes into the yard. The panel would be a reliable and inexpensive way to provide critical information about the cause of No Power faults.
>
> The panel would work in the following way. It would be mounted in Electric Locker #1, in which connection points are already available on the terminal boards. The circuit diagrams attached to this proposal (see Appendix B, page 14) show the necessary electrical modifications: the four points on the diagram labeled 1 through 4 are normally at zero volts DC.
>
> If the car's traction power drops below 95 amperes, point 1 will become 32 volts DC, starting the No Power Time Delay. After 25 seconds, the No Power Light will be energized and will stay lit even after power has been restored to 95 amperes. The 95-ampere Relay will then be found in its normal position, showing no problems, when checked by the yard mechanics. If any of the other three relays—Wheel Slips, Train Overspeed, or High-Voltage Failure—become energized by failures, the Power Dumping Relay will cut out propulsion power. The 95-ampere Relay will then supply 32 volts DC to the No Power Relay, which will again light the No Power Light after 25 seconds. The three relays will reset automatically, and again the yard mechanics will find all relays in their normal positions.
>
> With the installation of an electronic panel containing a latching driver circuit, the four points can be monitored electronically, and the LEDs on the panel will indicate which of the fault relays actually caused the lighting of the No Power light.

Figure 16–2 A Clear Plan of Action (Source: Schiller [1992])

Notice how the writer has provided a clear, logical explanation of his proposed idea, indicating how it would work and why it would probably solve an expensive problem. Later in the proposal he describes the need to train yard personnel in the proper use of the panel. In addition, he describes how the panel will be evaluated after an appropriate break-in period. He also provides specific cost data.

Demonstrate Your Professionalism and Credibility

The three journalistic questions that follow can help prompt you for the information that tells the reader that you (or your organization) carry through on your promises and that you have in place the necessary resources and management structures to ensure successful completion of the project.

Who

Who will be performing the work? If the project involves more than one person, you should describe the role of each important participant. In addition, you should make the case that they have the technical credentials and professionalism needed to complete the project successfully. Often you will provide résumés of the project leaders and describe similar projects that they have completed successfully. See Doyle and Berry (1993) for detailed advice on how to describe the key players.

In addition, you should provide a detailed management plan so your reader will understand that the project will be coordinated and overseen professionally. Thomas Ray (1993) recommends that you make sure you comply with all the management requirements spelled out in the RFP, include a checklist that cross-references the reader's needs and the relevant sections of your proposal, and describe specifically how you will organize and manage the project.

Where

Where will you perform the tasks that make up the project? Are your facilities—labs, equipment, etc.—adequate for the job? Often these questions are answered in an appendix.

When

When will you perform the various tasks involved in the project? Describe how any external factors, such as weather or personnel needs, will affect when the project can begin or how it will proceed. And describe the sequence of events that make up the project itself: what tasks must be completed before other tasks can begin and how long each task should take. Generally, these factors are described in an appendix in a graphical form. Gantt charts, or horizontal bar charts, are used commonly to show the duration and the temporal relationships of the various stages of the project. Other forms of charts, the critical path method (CPM) and program evaluation and review technique (PERT), help you show the logical sequencing of tasks: which tasks have to be completed before others can be started. All these types of charts are readily available on software programs.

In Figure 16–3, an excerpt from his proposal, the engineer describes his credentials for carrying out the project.

> I am presently working in the Maintenance Department as a Rail Vehicle Electrician 1st. Class. I have worked at Wayne Junction Car Shop and am now stationed at Powelton Avenue Yard. I have been certified in DC Motor Maintenance, Propulsion Circuits, and Auxiliary Controls, having successfully completed training programs at the Lenni Technical Center.
>
> For five years I was employed at NWL Transformers as a test technician, where my primary duty was to test power supplies that converted three-phase power to DC. As Quality Control Technician, I inspected the various analog control devices and safety interlocks on these supplies. On occasion I was sent on site to troubleshoot or start up supplies as a company representative for the Field Service Division.

Figure 16–3 A Statement of Credentials (Source: Schiller [1992])

Use Graphics to Help Your Readers See Your Ideas

Although graphics are critical in many kinds of workplace writing, they are absolutely fundamental to proposal writing. Because a proposal is a sales document, writers have developed strategies for packaging information in concise, easy-to-understand units that rely on graphics. Perhaps the best-known strategy for incorporating graphics is called STOP (sequential topical organization of proposals), a technique developed by Hughes Aircraft in 1962. In a STOP proposal, information is packaged in two-page spreads called topic modules. Figure 16–4 shows the basic elements of a topic module.

If one organization creates a technique called STOP, it's a good bet another will create one called GO. Graphics-oriented proposals, used originally at TRW, rely even more heavily on graphics; even the text is presented as a graphic, as shown in Figure 16–5. For detailed information on STOP, GO, and other approaches to combining text and graphics, see Tracey (1993).

Evaluate the Proposal Carefully Before You Send It Out

Like every other kind of workplace writing, proposals benefit from careful review. In fact, review is considered so important in most organizations that a special team of personnel, often called the Red Team, reviews each proposal before it is sent out. In some organizations, there is a standing Red Team, a group that does nothing but review proposals.

The Red Team consists of technical and managerial personnel who were *not* part of the proposal-writing team but who have sufficient expertise in the subject that they can spot problems such as omissions, inconsistencies, and unclear passages. The Red Team performs two basic tasks: it studies the proposal to determine how well it responds to the needs of the readers, and it performs mock reviews of the proposal, assigning points to each section, just as the actual readers will. Depending on the organization's resources, the proposal is revised and revised again until it is as good as it can be—or until the deadline has arrived.

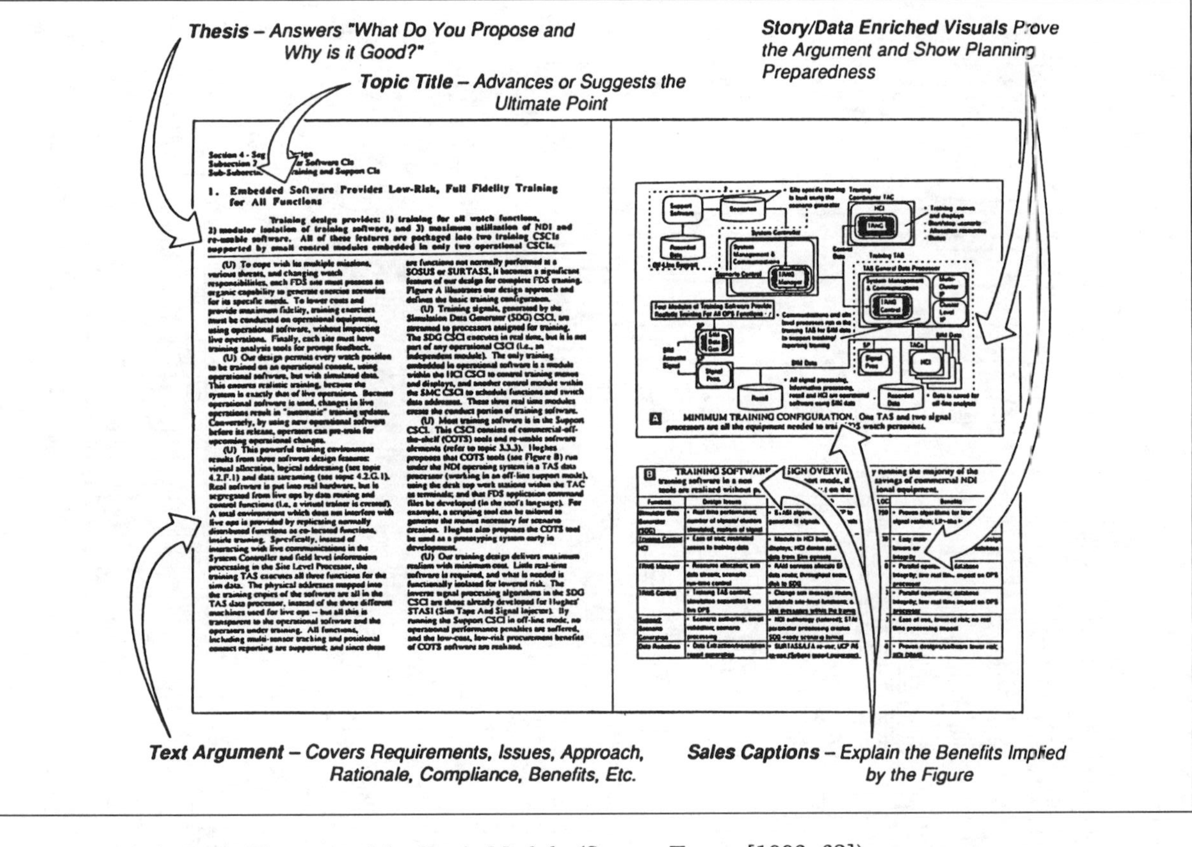

Figure 16–4 Basic Elements of the Topic Module (Source: Tracey [1993, 62])

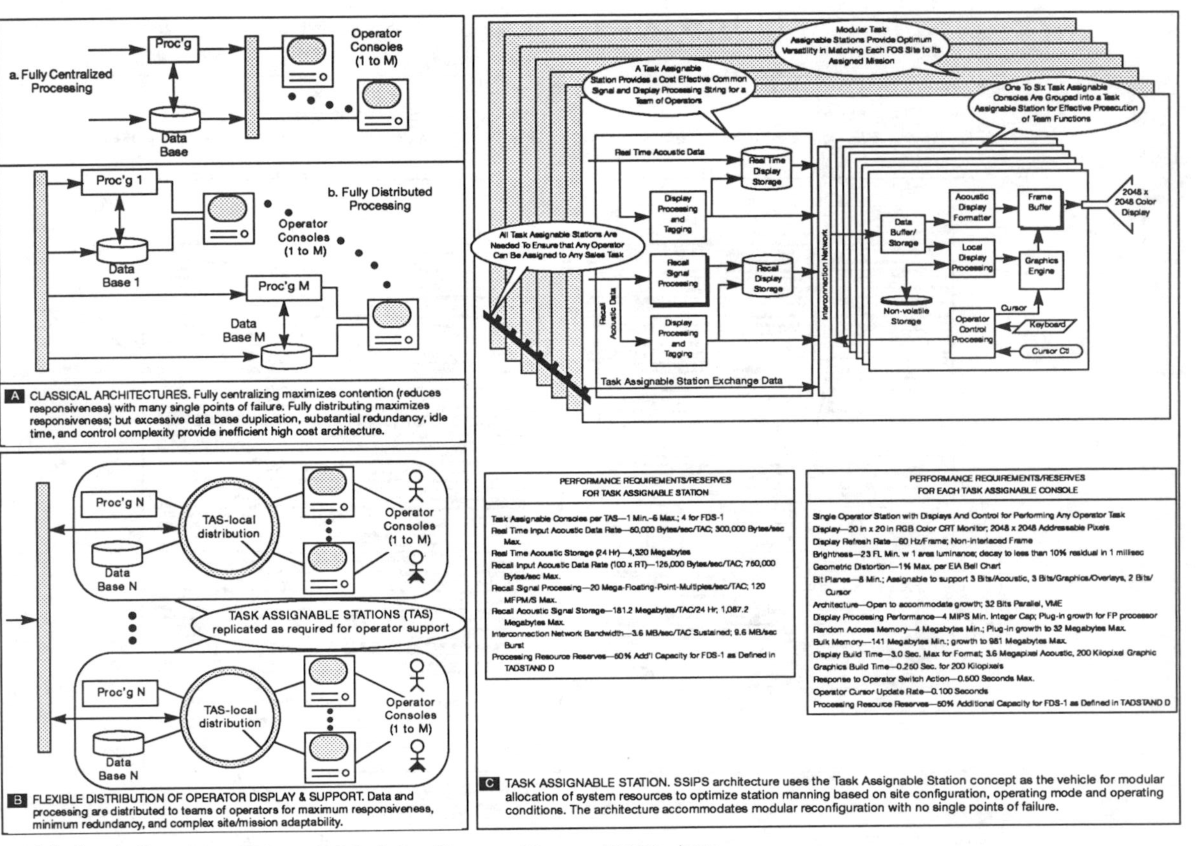

Figure 16–5 A Graphics-Oriented Module (Source: Tracey [1993, 72])

References

Doyle, M. L. and D. H. Berry. 1993. Presenting program personnel. In *How to create and present successful government proposals: Techniques for today's tough economy*, ed. J. W. Hill and T. Whalen, 181–188. New York: IEEE Press.

Hill, J. W. and T. Whalen, eds. 1993. *How to create and present successful government proposals: Techniques for today's tough economy*. New York: IEEE Press.

Holtz, H. 1984. *The consultant's guide to proposal writing: How to satisfy your clients and double your income.* New York: John Wiley.

Ray, T. E. 1993. Writing the management proposal. In *How to create and present successful government proposals: Techniques for today's tough economy*, ed. J. W. Hill and T. Whalen, 174–180. New York: IEEE Press.

Schiller, D. 1992. Proposal to modify the propulsion circuit on regional rail cars. Unpublished document.

Tracey, J. R. 1993. STOP, GO, and the state of the art in proposal writing. In *How to create and present successful government proposals: Techniques for today's tough economy*, ed. J. W. Hill and T. Whalen, 51–81. New York: IEEE Press.

Chapter 17

Progress Reports

A progress report, an update of a proposal, is a document in which you tell the readers how a project is proceeding and speculate on how it will conclude. For brief, inexpensive projects, there might be no progress report at all, or it might be a phone call, a brief memo, or a table listing tasks completed, tasks remaining, and comments. For a lengthy, expensive project, progress might be reported in a series of long, formal reports submitted monthly or quarterly.

The challenge of writing a good progress report is to communicate the information clearly and honestly.

Clarify the Context

A progress report is an opportunity for you to check in with your reader after the project has begun but before it is complete. Your first task in writing a progress report is to make sure your readers understand that the document *is* a progress report.

Isn't it obvious? It is to you, but it might not be obvious to your reader. The reason for this is that you as the writer are steeped in the project; it is likely to be your main task at work or even your only task. Your readers, however, might have a dozen other projects going and are likely to have forgotten the details of your project—or forgotten the project entirely. Therefore, you need to be extremely clear and specific in clarifying the context of the progress report. Make sure you communicate two pieces of information effectively:

- *The housekeeping details.* If you are using a memo format, put the phrase "progress report" in the subject heading; if you are using a report format, put that phrase in the title. Also indicate the number of the progress report and the intervals you are using: "This is the third monthly progress report in the high-definition TV screen display research project."
- *The background of the project.* To be safe, assume that your reader has forgotten the broad outlines of the project. Include any background discussions and the problem-definition statement from the original proposal. (The progress report is the kind of document that exploits the potential of the word processor, for you can copy the proposal and then update it.) If you use clear headings, your readers can skip over any sections they don't want to read.

Explain the Progress Clearly

The heart of the progress report is your answer to the readers' question, "What have you done for me lately?" and your speculation about future work on the project.

Two patterns of organization are used commonly:

Time

Perhaps the simplest pattern is based on the passage of time. In an outline form it looks like this:

1. work completed
2. future work

Some writers sandwich in a third category, "present work," which enables them to focus on the tasks they're working on at the moment. Other writers don't recognize a present tense; either the work is done, in which case it's in the past, or it isn't done, in which case it's in the future.

You can add subheadings based on the nature of the project. The most common approach calls for a task breakdown:

1. work completed
   1.1 task A
   1.2 task B
2. future work
   2.1 task A
   2.2 task B

The time pattern works well for relatively simple and short progress reports or for those that involve few different tasks.

Task

The task pattern is the time pattern turned inside out:

1. task A
   1.1 work completed
   1.2 future work
2. task B
   2.1 work completed
   2.2 future work

The task pattern works well with more complicated, longer progress reports, especially those with numerous tasks being carried out at the same time. In such cases, the organizational focus on the separate tasks results in a clearer presentation.

Explain the Progress Honestly

As I mentioned in Chapter 1, honesty is the chief characteristic of effective workplace writing. I want to mention honesty again at this point because progress reports pose a great temptation for a writer to lie or mislead.

The reason for this is simple: nobody likes to be the bearer of bad tidings. You don't want to disappoint your readers, and you don't want to get in trouble for not delivering what you promised in the proposal. So it is tempting to put a positive spin on the project and hope that in the remaining time you can make up for any problems. This kind of thinking isn't fully logical, of course, but even writers who have been disappointed a dozen times before often find it irresistible.

It's best to be honest for two reasons:

- *Ethical.* You owe it to your readers to be honest with them, just as you want them to be honest with you.
- *Practical.* If you're honest and things don't go as you hope, you can be accused of having been overly optimistic at the start of the project, of having been the victim of bad luck, or even of having made mistakes in doing your work. But none of these accusations is likely to prove fatal; usually, you are in serious trouble only when it is also discovered that you weren't straightforward in the progress report.

In thinking about honesty in progress reports, consider that the earlier you tell your readers that things aren't going according to plan, the better they will be able to accommodate the bad news. You are worrying about your own project, but they might well be trying to coordinate your project and half a dozen others. For this reason, you should be sure to be forthcoming about three important aspects of the project:

- *Time.* No matter how many projects you've worked on and how many times you've found yourself slipping even further behind schedule, it's only natural to hope that in *this* project you can make up the lost time. But if you lost 10 days in the first half of the project, it's more likely that you'll lose another 10 in the second half than that you'll make up the first 10.
- *Money.* This one is easy to talk about: if you suspect you are going to need extra money, the earlier you explain it to your readers, the more likely you are to get it.

- *Outcome.* If the final outcome of the project is not going to be what you promised, tell your readers as soon as possible.

The bottom line is that you don't want your reader to ask you why you didn't announce the bad news earlier.

Figure 17–1 is the body of the progress report by the engineer working for the rail system.

Summary:
The work is proceeding on schedule. I have chosen the four cars on which to install the panels and have built the panels themselves. I still have to install them in the cars. I expect the project to be completed, as specified in the proposal, on time.

Introduction:
This is the progress report on the first week's work on the project to build and install electronic panels on the General Electric cars to indicate the state of pertinent relays at the time of the No Power fault.

Work Completed:

- Task 1. Choosing the Four Cars on Which to Install the Panels
  To monitor the number of No Power faults occurring on General Electric rail cars, I have reviewed the Daily Hit Sheets for the past six months. The Daily Hit Sheets list in-service failures by car, and they report the time and type of failure. I wanted to know not only which cars had repeated No Power faults within the last six months, but also which cars were still listed as unrepaired by the Mechanical Department. Whenever a car is listed on the Daily Hit Sheet, it is checked out by yard mechanics. If they cannot find a defect, they use a No Defect Found code on the repair work order. From this information, I have chosen Regional Rail cars #9019, 326, 127, and 114 for modification. These four cars were failing at an average rate of every 15 days and therefore would be most likely to develop a No Power fault within the one-month trial period of the panel modification. This task, therefore, is complete.
- Task 2. Building and Installing the Electronic Panels
  I have completed building the electronic panel to indicate the state of pertinent relays at the time of the No Power fault. This panel will hold information in memory to be used to determine

the actual problem that occurred on the car. Bench tests on the panels indicate that they are working according to specifications. The first half of this task, therefore, is complete.

Future Work:

- Task 2. Building and Installing the Electronic Panels
  Now that the electronic panels have checked out satisfactorily, I have to install them in the four cars and check them again. This portion of the task should be completed by the end of this week.

Conclusion:
The project is proceeding on schedule, without any unanticipated problems. I expect the project and its accompanying report to be completed on schedule.

Figure 17–1 Body of Progress Report (Source: Schiller [1992])

Reference

Schiller, D. 1992. Progress report on the project to modify the propulsion circuit on regional rail cars. Unpublished document.

Chapter 18

Completion Reports

This chapter discusses completion reports, which are often called *final reports* or just *reports.* A completion report is the culmination of a substantial research project. Although every completion report is different, two basic types exist:

- *Physical research reports* communicate information about a research project that took place in the lab or in the field. A consulting engineering team might have designed and installed a new scrubber for an industrial plant or assessed storm damage to a building, or a botanist might have conducted an experiment to determine the effect of a new chemical to retard the growth rate of an agricultural pest.
- *Feasibility studies* report on which course of action an organization should take. For instance, a company might need to decide whether to create its own R&D department or continue to subcontract its R&D work. A feasibility study on this topic

would present the relevant facts, draw conclusions, and offer recommendations on which course of action is wiser.

Although there are many variations on these two kinds of reports, they share a basic strategy: the writer introduces the readers to the report, describes the working methods followed in the project, presents the data created or discovered, draws conclusions from these data, and, in some cases, offers recommendations on the basis of these conclusions.

This basic strategy is displayed in the body of the report. Often, completion reports also contain many of the formal elements described in Chapter 15.

Introduce the Report Clearly

Chapter 5 explains that effective introductions often describe six aspects of the document that follows:

- the subject
- the purpose
- the background
- the scope
- the organization
- the key terms

These six aspects apply to most reporting situations. Keep in mind that you probably have already written the first four of the six—the subject, purpose, background, and scope—in the proposal and in any progress reports and that you might have written the other two as well. The completion report, like the progress report, is a word-processed document; you copy whole chunks from the previous document. If the project went smoothly, the only changes you will have to make to the introduction are to substitute the past tense for the future tense in some places. Whereas earlier you wrote that the project *will study* the feasibility of entering the DRAM market, now you write that the project *studied* the feasibility.

Figure 18–1 is an excerpt from the introduction to the report written by the engineer working for the regional rail line (see Chapter 16 for excerpts from his proposal). The writer has already described the background and the problem. Here he addresses the scope of the report.

> This report describes the project to build and install these panels on four regional rail cars and then to monitor the effectiveness of the panels over a one-month period. After describing the methods used, I report the results: the failures on the four cars and the resulting actions taken by shop maintenance personnel. (Appendix D, page 19, shows the full data on the failures and the maintenance hours logged in servicing the cars.) In addition, I provide my conclusions on the effectiveness of the panels and my recommendations for further implementation. Cost data are included in Appendix E, page 21.

Figure 18–1 Excerpt from a Report Introduction (Source: Schiller [1992])

Describe Your Methods

In virtually every report, you must report the methods you used in carrying out the project. Depending on the field you are working in, however, this description can range from brief and general to lengthy and precise.

For instance, the methods section of a report might consist of a sentence or two, in which you merely state that you used Pearson's correlations to run the statistics; the readers will assume that you used them correctly. Some cases, however, call for a justification of your methodology. As discussed in Chapter 16 on proposals, you might need to defend your choice of Pearson's correlations as the appropriate test. What was it about the nature of the data that called for that test? If your readers don't know which test to run or if they would expect you to have run a different test, a full defense of your choice might be appropriate, because if they don't understand or accept your methods, the report ultimately will fail.

If you are writing a feasibility study, be sure to make clear in the methods section how you will be evaluating the options. Just as you don't go to a car dealer without having thought carefully about what kind of car you need, you do not evaluate options without considering what criteria you will be using—and how you will weigh them—in your analysis. If you will be using a decision matrix to assign a certain point value to each criterion you study, describe that decision matrix and show why it is the appropriate one to use for this kind of study.

Figure 18–2 is an excerpt from the methods section of the report by the engineer working for the regional rail line. Here he describes how he chose the four cars on which to install the panels.

Choosing the Four Cars on Which to Install the Panels

To monitor the number of No Power faults occurring on General Electric rail cars, I reviewed the Daily Hit Sheets for the past six months. The Daily Hit Sheets list in-service failures by car, and they report the time and type of failure. I wanted to know not only which cars had repeated No Power faults within the last six months, but also which cars were still listed as unrepaired by the Mechanical Department. Whenever a car is listed on the Daily Hit Sheet, it is checked out by yard mechanics. If they cannot find a defect, they use a No Defect Found code on the repair work order. From this information, I chose Regional Rail cars #9019, 326, 127, and 114 for modification. These four cars were failing at an average rate of every 15 days and therefore would be most likely to develop a No Power fault within the one-month trial period of the panel modification. (See Appendix A, page 13, for the data on the No Power faults over the last six months on the four cars, and Appendix B, page 15, for the No Power faults for the same period for the entire fleet of cars.)

Figure 18–2 Excerpt from a Methods Section (Source: Schiller [1992])

Present Your Results

Results are the data you collect or generate. When you present your data, you have to decide how much to include in the body of the report and how much to relegate to an appendix. As discussed in Chapter 15, the body is the appropriate location if you want your readers to study them; an appendix is the appropriate location if you merely want to make them available but do not think your readers need to study them to understand the report. In most cases, it is best to present the most important results in the body and the full results in an appendix, making sure to cross-reference the readers to the appendix.

Figure 18–3 is an excerpt from the results section of the regional rail report.

> *Car #114*
>
> During the test period, car #114 experienced a High-Voltage Failure, which was diagnosed effectively because of the panel.

Figure 18–3 Excerpt from a Results Section (Source: Schiller [1992])

Notice that this discussion is brief; at this point the writer wants to provide only a brief summary. In an appendix, however, the writer provided the discussion shown in Figure 18–4.

> *Car #114*
>
> During the test period, car #114 experienced a High-Voltage Failure, which was diagnosed effectively because of the panel.
>
> The car developed a No Power fault while in service on the Ivy Ridge line. The car was brought to Powelton Avenue Yard for service at 9:45 p.m., February 10. The Mechanical Department Yard Foreman notified me, and I was able to go to the yard in time to assist the Mechanical Department with the troubleshooting. We found that the panel hat latched out, indicating a High-Voltage Failure. The High-Voltage Failure is designed to cut out traction power if the overhead catenary wire voltage drops below 8,500 volts. Since car #114 was the only one in a five-car train to develop a No Power fault, the catenary voltage probably did not drop to 8,500 volts. In a thorough check of the High-Voltage Failure circuit, we found a loose wire on resistor HV-28. This is one of the four step-down resistors that feed the High-Voltage Failure relay diodes. Apparently the wire was tight enough to keep the High-Voltage Failure relay in its normal state most of the time, but occasionally the wire opened the circuit to the relay, dropping out the High-Voltage Failure relay.
>
> In discussing this problem with the Mechanical Department personnel, I learned that the High-Voltage Failure circuit is never

checked on a car that is reported for a No Power fault. From this we concluded that car #114 would not have been repaired properly had the panel not been installed.

Figure 18–4 Excerpt from an Appendix (Source: Schiller [1992])

Draw Conclusions from the Results

Conclusions are the inferences you draw from the results. In other words, conclusions explain the results. For instance, if the EPA allows you to produce 3 ppb of CO in your effluent and the results of your analysis show that you produce 4 ppb, the appropriate conclusion is that you are in violation of the EPA statute. If the project were to determine which brand of oscilloscope to purchase for your lab, your conclusion might be that Brand A is the best one but that Brand B is the best value.

As is the case in writing journal articles, it is best to present the results and conclusions separately. Even though in doing the project you do not progress mechanically from data gathering to data analysis, the separation in the report makes it easier for your reader to follow your thinking.

Figure 18–5 is the conclusion from the regional-rail report.

6. Conclusion

Eighty-three cars had No Power faults but were not effectively repaired during my six-month monitoring of the Daily Hit Sheet. The electronic panel was successfully used to help repair the three cars that had No Power faults repeatedly during that six-month period. The panel saved the Mechanical Department personnel some 48 hours repairing the cars during the one-month trial period.

Adding the panel to all 234 remaining cars would cost $50,427 but would save approximately $35,700 per year, for a payback time of 1.4 years. (See Appendix E, page 21, for the full financial data.) On the basis of these data, I conclude that adding the panels to all the cars would reduce our maintenance costs considerably and improve the quality of the service we provide.

Figure 18–5 Conclusion Section (Source: Schiller [1992])

Present Your Recommendations

Not all reports end with recommendations, but most do. Recommendations are usually presented separately from conclusions for the same reason that conclusions are presented separately from results: the separation makes it easier for your readers to follow your thinking.

Although the recommendations often follow logically and obviously from results, sometimes you have to do some explaining. Here is an example of a simple situation: in a feasibility study of three different methods of carrying out a task, your conclusion is that Method A is most effective and most efficient. If, according to the criteria you set up, Method A is clearly the best, it makes sense to recommend that your company pursue Method A. Here is an example of a more common situation: Method A is the most effective, but Method B is more efficient. If money were no problem, you would recommend Method A, but Method B appears to be a better deal. In this case, your recommendation might be to pursue Method A if the budget can support it but to pursue Method B if we can't afford Method A, or it might be to do further study to see whether it is possible to bring down the price on Method A or improve the capabilities of Method B.

And don't forget one sort of conclusion that often makes the best sense: don't do anything. People sometimes mistakenly think that a research project has to call for a dramatic recommendation at the end. Beware of a solution that looks too good to be true, because, as the old saying goes, it probably is. The last thing you want to do is recommend a course of action, only to find out too late that you overlooked some problems with it.

Figure 18–6 is the recommendation from the regional-rail report.

7. Recommendation

I recommend that the panel modification be performed as soon as possible on all General Electric cars. Because the panels can be installed in less than four hours per car, this project would require no disruption of our rail schedule. In addition, this project would require no additional hiring of maintenance personnel, because the maintenance workload would be reduced with the installation of each panel.

Figure 18–6 Recommendation Section (Source: Schiller [1992])

Reference

Schiller, D. 1992. Propulsion power circuit modification on regional rail cars: A recommendation. Unpublished document.

Appendices

Appendix A

Checklists

Following are 16 checklists that highlight the material covered in this book. Each item is followed by a page reference that directs you to the appropriate discussion in the book.

As you look at these checklists, remember that every writer is different, and therefore every writer's checklists should be different. I might have listed items that you never have any problems with, and I might have omitted an item that would be helpful. Therefore, you should modify these checklists so that they work effectively for you.

Checklist 1. The Writing Process

1. Have you analyzed four aspects of your readers?

a. their professional characteristics *see page 17*
b. their personal characteristics *see page 17*

c. their attitudes toward the subject *see page 17*
d. their reasons for reading the document *see page 18*

2. Have you defined your purpose in a sentence so that you have a clear direction to follow in the document? *see page 18*
3. Have you evaluated the implications of your audience and purpose? *see page 19*
4. Have you made sure your boss agrees with what you've decided by outlining your strategy in a memo? *see page 21*
5. Have you generated ideas to include in the document by brainstorming, talking, free writing, or sketching? *see page 23*
6. Have you organized your information by stating your overall purpose and creating logical categories? *see page 26*
7. Have you sequenced the ideas within the categories using such patterns as chronological, spatial, classification, partition, general to specific, more important to less important, problem-methods-solution, and cause-effect? *see page 29*
8. Have you written the draft fast, without stopping to revise? *see page 34*
9. Have you let the draft sit as long as possible so that you can forget it? *see page 35*
10. Have you read the draft aloud to hear what it sounds like? *see page 35*
11. Have you gotten help from someone else? *see page 35*
12. Have you reviewed the document, looking for problems in areas such as comprehensiveness, accuracy, organization, emphasis, paragraphing, style, and spelling? *see page 35*

Checklist 2. Using the Computer to Improve Your Writing

1. Have you tried brainstorming at the computer? *see page 40*
2. Have you tried drafting right on the outline? *see page 41*
3. Have you tried invisible writing? *see page 42*
4. Have you tried abbreviating using the search-and-replace function? *see page 42*
5. Have you tried using the computer in collaborating? *see page 42*

6. Have you tried using the block-copy function to try out different organizational patterns in your document? *see page 43*
7. Have you used a spell checker (and then reread the document carefully to make sure you are spelling the right word correctly)? *see page 43*
8. Have you used a thesaurus program (and made sure you understand the connotations of any word you are choosing as a synonym)? *see page 44*
9. Have you used a style program (and made sure the advice is appropriate before deciding to take it)? *see page 44*

Checklist 3. Improving Coherence in Your Writing

1. Do your titles and headings clearly indicate the subject and purpose of the document? *see page 48*
2. Are your titles and headings precise, easy to read and understand, and structured appropriately for the subject and the audience? *see page 49*
3. Is parallel information presented in a list format, with the lead-in indicating the number of items in the list? *see page 50*
4. Does your introduction indicate the subject, purpose, background, scope, and organization of the document as well as the key terms you use? *see page 52*
5. Do your conclusions reiterate the main points of the discussion, recommend what should be done next, help the reader find out more information, and offer your services in the future? *see page 55*

Checklist 4. Paragraphing

1. Does each paragraph begin with a clear topic sentence? *see page 59*
2. Is the topic sentence supported logically? *see page 61*
3. Is the coherence of the paragraphs emphasized by the use of key terms, transitional words and phrases, and demonstrative pronouns—*this, that, these,* and *those*—followed by nouns? *see page 62*

4. Are the paragraphs short enough so that they are easy to read and understand? *see page 65*

Checklist 5. Sentences

1. Does the document follow the appropriate stylistic guidelines? *see page 68*
2. Are the active and passive voices used appropriately? *see page 68*
3. Are your sentence patterns appropriate for the subject, audience, and purpose? *see page 70*
4. Have you focused on the real subject of the sentence, without burying it in a prepositional phrase or expletive? *see page 72*
5. Have you focused on the real verb, without using inappropriate nominalizations? *see page 73*
6. Are your restrictive and nonrestrictive modifiers used correctly? *see page 74*
7. Have you avoided misplaced modifiers and dangling modifiers? *see page 75*
8. Is parallel information presented in parallel forms? *see page 76*

Checklist 6. Words and Phrases

1. Does the document use simple, clear words and phrases? *see page 79*
2. Does the document avoid unnecessary jargon? *see page 82*
3. Does the document avoid euphemisms? *see page 82*
4. Does the document avoid clichés? *see page 82*
5. Does the document avoid sexist language? *see page 83*
6. Have you used readability formulas carefully? *see page 84*

Checklist 7. Graphics

1. Have you determined whether you need graphics? *see page 87*
2. Have you determined what kinds of graphics would best communicate the points you wish to make? *see page 89*

3. Have you made sure that the graphics accurately and honestly represent your point? *see page 90*
4. Are the graphics self-sufficient? *see page 91*
5. Are the graphics placed in appropriate locations? *see page 95*
6. Are the graphics effectively introduced and their significance explained? *see page 95*

Checklist 8. Page Design

1. Are the margins adequate for the difficulty of the material and the size of the page? *see page 100*
2. Have you considered a multicolumn format to save space, increase readability, and integrate text and graphics effectively? *see page 101*
3. Is the line spacing within the body text and between sections appropriate? *see page 102*
4. Is the justification appropriate? *see page 104*
5. Are the typefaces appropriate? *see page 106*
6. Have you taken advantage of the different members of the type family? *see page 107*
7. Are the type sizes varied appropriately? *see page 108*
8. Are both upper- and lowercase type used appropriately? *see page 110*
9. Have you designed the titles and headings for clarity by using size and indentation effectively? *see page 110*
10. Have you designed the lists for clarity by using upper- and lowercase, indentation, and end punctuation effectively? *see page 112*

Checklist 9. Letters

1. Does your letter use a conventional format? *see page 117*
2. Do your salutation and complimentary close follow protocol? *see page 118*
3. Does your first paragraph link your letter to the most recent communication with your reader and provide an overview of the rest of the letter? *see page 122*

4. Does your letter end on a positive note? *see page 123*
5. Does your vocabulary sound natural? *see page 124*

Checklist 10. Memos

1. Does the memo begin with an informative to-from-subject-date heading? *see page 128*
2. Does the body of the memo begin with a clear statement of your purpose? *see page 128*
3. Have you included a summary for memos of more than one page? *see page 129*
4. Does the memo conclude with an action statement? *see page 130*

Checklist 11. Minutes

1. Do the minutes include the housekeeping details, including information about the time and place and the attendees? *see page 136*
2. Do the minutes record events accurately? *see page 136*
3. Do the minutes record events diplomatically? *see page 137*

Checklist 12. Procedures and Manuals

1. Have you emphasized safety in the procedure or manual by writing and designing the safety warnings clearly and placing them in appropriate locations? *see page 142*
2. Have you provided preliminary information that orients the reader, including a how-to-use-this-manual section? *see page 145*
3. Are the instructions clearly written? *see page 148*
4. Are graphics included to clarify the text? *see page 150*
5. Have you designed the document for easy use in the field, considering such factors as its size, the type of paper and binding, and the design of the pages? *see page 152*
6. Have you anticipated the need for updates by thinking about the type of binding and the style of pagination to use? *see page 154*

7. Are troubleshooting tips included? *see page 154*

Checklist 13. Formal Elements of Reports

1. Is a transmittal letter included? *see page 157*
2. Does your title page clearly and emphatically display the title and indicate the date of submission and the names of the writer and the principal reader? *see page 160*
3. Is your table of contents clear and specific, and does it reflect accurately the headings in the document itself? *see page 160*
4. Is a list of illustrations included, if appropriate? *see page 162*
5. Have you included an informative or descriptive abstract, as appropriate? *see page 162*
6. Is an executive summary included for the convenience of your managerial readers? *see page 164*
7. Are appendices included for information that most of your readers will not read? *see page 166*

Checklist 14. Proposals

1. Does the introduction show that you understand the readers' needs? *see page 171*
2. Do you propose a clear, specific technical plan for the project? *see page 174*
3. Have you demonstrated your professionalism and credibility? *see page 176*
4. Have you included graphics to clarify your ideas? *see page 177*
5. Have you reviewed the proposal carefully before sending it out? *see page 178*

Checklist 15. Progress Reports

1. Is the context of the progress report made clear? *see page 183*
2. Is your progress explained clearly? *see page 184*
3. Is your progress explained honestly? *see page 185*

Checklist 16. Completion Reports

1. Is the report introduced clearly? *see page 190*
2. Are your methods described clearly? *see page 191*
3. Are your results presented clearly? *see page 192*
4. Do your conclusions derive clearly from the results? *see page 194*
5. Do your recommendations derive clearly from the conclusions? *see page 195*

Appendix B

Handbook

This handbook concentrates on style, punctuation, and mechanics. Where appropriate, it defines common errors directly after discussing the correct usage.

Many of the usage recommendations made here are only suggestions. If your organization or professional field has a style guide that makes different recommendations, you should, of course, follow it.

Sentence Style

Avoid Sentence Fragments

A sentence fragment is an incomplete sentence. Most sentence fragments are caused by one of two problems:

- a missing verb

Fragment: The pressure loss caused by a worn gasket.
Complete sentence: The pressure loss was caused by a worn gasket.
Complete sentence: The pressure loss caused by a worn gasket was identified and fixed.

Fragment: A 486-series computer equipped with a VGA monitor.
Complete sentence: It is a 486-series computer equipped with a VGA monitor.
Complete sentence: A 486 computer equipped with a VGA monitor will be delivered today.

- a dependent element used without an independent clause

Fragment: Because the data could not be verified.
Complete sentence: Because the data could not be verified, the article was not accepted for publication.
Complete sentence: The article was not accepted for publication because the data could not be verified.

Fragment: For a given rectangular waveguide.
Complete sentence: For a given rectangular waveguide, the *a* and *b* dimensions are fixed.
Complete sentence: The *a* and *b* dimensions are fixed for a given rectangular waveguide.

Avoid Comma Splices

A comma splice is the error in which two independent clauses are joined, or spliced together, by a comma. Independent clauses can be linked correctly in three different ways:

- by a comma and a coordinating conjunction

Comma splice: The 909 printer is our most popular model, it offers an unequaled blend of power and versatility.

Correct: The 909 printer is our most popular model, for it offers an unequaled blend of power and versatility.

In this case, a comma and one of the coordinating conjunctions (*and, or, not, but, for, so,* and *yet*) link the two independent clauses. The coordinating conjunction explicitly states the relationship between the two clauses.

- by a semicolon

Comma splice: The 909 printer is our most popular model, it offers an unequaled blend of power and versatility.
Correct: The 909 printer is our most popular model; it offers an unequaled blend of power and versatility.

In this case, a semicolon is used to link the two independent clauses. The semicolon creates a somewhat more distant relationship between the two clauses than the comma-and-coordinating conjunction link; the link remains implicit.

- by a period or other terminal punctuation

Comma splice: The 909 printer is our most popular model, it offers an unequaled blend of power and versatility.
Correct: The 909 printer is our most popular model. It offers an unequaled blend of power and versatility.

In this case, the two independent clauses are separate sentences. Of the three ways to punctuate the two clauses correctly, this punctuation suggests the most distant relationship between them.

Avoid Run-On Sentences

A run-on sentence (sometimes called a fused sentence) is a comma splice without the comma. In other words, two independent clauses appear without any punctuation between them. Any of the three strategies for fixing commas splices fixes run-on sentences.

Run-on sentence: The 909 printer is our most popular model it offers an unequaled blend of power and versatility.

Correct: The 909 printer is our most popular model, for it offers an unequaled blend of power and versatility.

Correct: The 909 printer is our most popular model; it offers an unequaled blend of power and versatility.

Correct: The 909 printer is our most popular model. It offers an unequaled blend of power and versatility.

Avoid Ambiguous Pronoun References

Pronouns must refer clearly to the words or phrases they replace. Ambiguous pronoun references can lurk in even the most innocent-looking sentences:

Unclear: Remove the cell cluster from the medium and analyze it. (*Analyze what, the cell cluster or the medium?*)

Clear: Analyze the cell cluster after removing it from the medium.

Clear: Analyze the medium after removing the cell cluster from it.

Clear: Remove the cell cluster from the medium. Then analyze the cell cluster.

Clear: Remove the cell cluster from the medium. Then analyze the medium.

Ambiguous references can also occur when a relative pronoun such as *which* or a subordinating conjunction such as *where* is used to introduce a dependent clause:

Unclear: She decided to evaluate the program, which would take five months. (*What would take five months, the program or the evaluation?*)

Clear: She decided to evaluate the program, a process that would take five months. (*By replacing "which" with "a process that," the writer clearly indicates that the evaluation will take five months.*)

Clear: She decided to evaluate the five-month program. (*By using the adjective "five-month," the writer clearly indicates that the program will take five months.*)

Unclear: This procedure will increase the handling of toxic materials outside the plant, where adequate safety measures can be taken.

(*Where can adequate safety measures be taken, inside the plant or outside?*)

Clear: This procedure will increase the handling of toxic materials outside the plant. Because adequate safety measures can be taken only in the plant, the procedure poses risks.

Clear: This procedure will increase the handling of toxic materials outside the plant. Because adequate safety measures can be taken only outside the plant, the procedure will decrease safety risks.

As the last example shows, sometimes the best way to clarify an unclear pronoun is to split the sentence in two, eliminate the problem, and add clarifying information. Clarity is always an important characteristic of workplace writing. If more words will make your writing clearer, use them.

Ambiguity can also occur at the beginning of a sentence:

Unclear: Allophanate linkages are among the most important structural components of polyurethane elastomers. They act as cross-linking sites. (*What act as cross-linking sites, allophanate linkages or polyurethane elastomers?*)

Clear: Allophanate linkages, which are among the most important structural components of polyurethane elastomers, act as cross-linking sites. (*The writer has changed the second sentence into a clear nonrestrictive modifier.*)

Your job is to use whichever means—restructuring the sentence or dividing it in two—will best ensure that the reader will know exactly which word or phrase the pronoun is replacing.

If you use a pronoun to begin a sentence, be sure to follow it immediately with a noun that clarifies the reference. Otherwise, the reader might be confused.

Unclear: The new regulations require that all researchers submit individual human-subjects research consent forms. These are scheduled to be discussed at the next senate meeting. (*What are scheduled to be discussed, the regulations or the forms?*)

Clear: The new regulations require that all researchers submit individual human-subjects research consent forms. These regulations are scheduled to be discussed at the next senate meeting.

Compare Items Clearly

When comparing or contrasting items, make sure your sentence clearly communicates the relationship. A simple comparison between two items often causes no problems: "The X3000 has more storage than the X2500." However, don't let your reader confuse a comparison and a simple statement of fact. For example, in the sentence "Trout eat more than minnows," does the writer mean that trout don't restrict their diet to minnows or that trout eat more than minnows eat? If a comparison is intended, a second verb should be used: "Trout eat more than minnows do." And if three items are introduced, make sure that the reader can tell which two are being compared.

Unclear: Trout eat more algae than minnows.
Clear: Trout eat more algae than they do minnows.
Clear: Trout eat more algae than minnows do.

Beware of comparisons in which different aspects of the two items are compared:

Illogical: The resistance of the copper wiring is lower than the tin wiring.
Logical: The resistance of the copper wiring is lower than that of the tin wiring.

In the illogical construction, the writer contrasts "resistance" with "tin wiring" rather than the resistance of copper with the resistance of tin. In the revision, the pronoun *that* is used to substitute for the repetition of "resistance."

Use Adjectives Clearly

In general, adjectives are placed before the nouns they modify: "the plastic washer." Workplace writing, however, often requires clusters of adjectives. To prevent confusion, use commas to separate coordinate adjectives, and use hyphens to link compound adjectives.

Adjectives that describe different aspects of the same noun are known as coordinate adjectives:

portable, programmable CD-ROM player
adjustable, step-in bindings

In this case, the comma replaces the word *and*.

Note that sometimes an adjective is considered part of the noun it describes: "electric drill." When one adjective is added to "electric drill," no comma is required: "a reversible electric drill." The addition of two or more adjectives, however, creates the traditional coordinate construction: "a two-speed, reversible electric drill."

The phrase "two-speed" is an example of a compound adjective—one made up of two or more words. Use hyphens to link the elements in compound adjectives that precede nouns:

a variable-angle accessory
increased cost-of-living raises

The hyphens in the second example prevent the reader from momentarily misinterpreting "increased" as an adjective modifying "cost" and "living" as a participle modifying "raises."

A long string of compound adjectives can be confusing even if hyphens are used appropriately. To ensure clarity in such a case, put the adjectives into a clause or phrase following the noun:

Unclear: an operator-initiated, default-prevention technique
Clear: a technique initiated by the operator for preventing default

Maintain Number Agreement

Number disagreement commonly takes one of two forms in workplace writing: (1) the verb disagrees in number with the subject when a prepositional phrase intervenes or (2) the pronoun disagrees in number with its referent when the latter is a collective noun.

Subject-verb disagreement A prepositional phrase does not affect the number of the subject and the verb. The following examples show that the object of the preposition can be plural in a singular sentence or singular in a plural sentence. (The subjects and verbs are italicized.)

Incorrect: The *result* of the tests *are* promising.
Correct: The *result* of the tests *is* promising.
Incorrect: The *results* of the test *is* promising.
Correct: The *results* of the test *are* promising.

Don't be misled by the fact that the object of the preposition and the verb don't sound natural together, as in *tests is* or *test are.* Grammatical agreement of subject and verb is the important consideration.

Pronoun-referent disagreement The problem of pronoun-referent disagreement occurs most often when the referent is a collective noun—one that can be interpreted as either singular or plural, depending on its usage:

Incorrect: The *company* is proud to introduce a new LAN configuration for *their* customers.
Correct: The *company* is proud to introduce a new LAN configuration for *its* customers.

In this example, "the company" acts as a single unit; therefore, the singular verb, followed by a singular pronoun, is appropriate. When the individual members of a collective noun are stressed, however, plural pronouns and verbs are appropriate: "The inspection team have prepared their reports." Or, "The members of inspection team have prepared their reports."

Punctuation

The Period

Periods are used in the following instances.

1. at the end of sentences that do not ask questions or express strong emotion

The lateral stress still needs to be calculated.

2. after some abbreviations

M.D.
U.S.A.

(For a further discussion of abbreviations, see pp. 237–239.)

3. with decimal fractions

4.056
$6.75
75.6%

The Exclamation Point

The exclamation point is used at the end of a sentence that expresses strong emotion, such as surprise or doubt:

> The nuclear plant, which was originally expected to cost $1.6 billion, eventually cost more than $8 billion!

Because workplace writing requires objectivity and a calm, understated tone, exclamation points are rarely used.

The Question Mark

The question mark is used at the end of a sentence that asks a direct question:

What did the commission say about effluents?

Do not use a question mark at the end of a sentence that asks an indirect question:

> He wanted to know whether the procedure had been approved for use.

When a question mark is used within quotation marks, the quoted material needs no other end punctuation:

"What did the commission say about effluents?" she asked.

The Comma

The comma is the most frequently used punctuation mark, as well as the one about whose usage authorities most often disagree. Following are the basic uses of the comma.

- to separate the clauses of a compound sentence (one composed of two or more independent clauses) linked by a coordinating conjunction (*and, or, nor, but, so, for, yet*)

  Both methods are acceptable, but we have found that the Simpson procedure gives better results.

  In many compound sentences, the comma is needed to prevent the reader from mistaking the subject of the second clause for an object of the verb in the first clause:

  The RESET command affects the field access, and the SEARCH command affects the filing arrangement.

  Without the comma, the reader is likely to interpret the coordinating conjunction "and" as a simple conjunction linking "field access" and "SEARCH command."

- to separate items in a series composed of three or more elements

  The manager of spare parts is responsible for ordering, stocking, and disbursing all spare parts for the entire plant.

The comma following the second-to-last item is required by most technical-writing style manuals, despite the presence of the conjunction "and." The comma clarifies the separation and prevents misreading. For example, sometimes in workplace writing the second-to-last item will be a compound noun containing an "and."

The report will be distributed to Operations, Research and Development, and Accounting.

- to separate introductory words, phrases, and clauses from the main clause of the sentence

However, we will have to calculate the effect of the wind.
To facilitate trade, government holds a yearly international conference.
Whether the workers like it or not, the managers have decided not to try the flextime plan.

In each of these three examples, the comma helps the reader follow the sentence. Notice in the following example how the comma actually prevents misreading:

Just as we finished eating, the rats discovered the treadmill.

The comma is optional if the introductory text is brief and cannot be misread.

Correct: First, let's take care of the introductions.
Correct: First let's take care of the introductions.

- to separate the main clause from a dependent clause

The advertising campaign was canceled, although most of the executive council saw nothing wrong with it.

Most accountants wear suits, whereas few engineers do.

- to separate nonrestrictive modifiers (parenthetical clarifications) from the rest of the sentence

Jones, the temporary chairman, called the meeting to order.

- to separate interjections and transitional elements from the rest of the sentence

Yes, I admit your findings are correct.
Their plans, however, have great potential.

- to separate coordinate adjectives

The heavy, awkward trains are still being used.

The comma here takes the place of the conjunction "and." If the adjectives are not coordinate—that is, if one of the adjectives modifies the combination of the adjective and the noun—do not use a comma:

They decided to go to the first general meeting.

- to signal that a word or phrase has been omitted from an elliptical expression

Smithers is in charge of the transportation; Harlen, the data management; Demarest, the publicity.

In this example, the commas after "Harlen" and "Demarest" show that the phrase "is in charge of" has been omitted.

- to separate a proper noun from the rest of the sentence in direct address

John, have you seen the purchase order from United?
What I'd like to know, Betty, is why we didn't see this problem coming.

- to introduce most quotations

He asked, "What time were they expected?"

- to separate towns, states, and countries

Bethlehem, Pennsylvania, is the home of Lehigh University.
He attended Lehigh University in Bethlehem, Pennsylvania, and the University of California at Berkeley.

Note the use of the comma after "Pennsylvania."

- to set off the year in dates

August 1, 1995, is the anticipated completion date.

Note the use of the comma after "1995." If the month separates the date from the year, the commas are not used, because the numbers are not next to each other:

The anticipated completion date is 1 August 1995.

- To clarify numbers

12,013,104

(European practice is to reverse the use of commas and periods in writing numbers: periods are used to signify thousands, and commas to signify decimals.)

- to separate names from professional or academic titles

Harold Clayton, Ph.D.
Joyce Carnone, P.E.

Note that the comma also follows the title in a sentence:

Harold Clayton, Ph.D., is the featured speaker.

Common Errors

- no comma between the clauses of a compound sentence

Incorrect: The mixture was prepared from the two premixes and the remaining ingredients were then combined.
Correct: The mixture was prepared from the two premixes, and the remaining ingredients were then combined.

- no comma (or just one comma) to set off a nonrestrictive modifier

Incorrect: The phone line, which was installed two weeks ago had to be disconnected.
Correct: The phone line, which was installed two weeks ago, had to be disconnected.

- no comma separating introductory words, phrases, or clauses from the main clause, when misreading can occur

Incorrect: As President Canfield has been a great success.
Correct: As President, Canfield has been a great success.

- no comma (or just one comma) to set off an interjection or a transitional element

Incorrect: Our new statistician, however used to work for Konaire, Inc.
Correct: Our new statistician, however, used to work for Konaire, Inc.

- comma splice (a comma used to "splice together" independent clauses not linked by a coordinating conjunction)

Incorrect: All the motors were cleaned and dried after the water had entered, had they not been, additional damage would have occurred.
Correct: All the motors were cleaned and dried after the water had entered; had they not been, additional damage would have occurred.
Correct: All the motors were cleaned and dried after the water had entered. Had they not been, additional damage would have occurred.

For more information on comma splices, see p. 209.

- superfluous commas

Incorrect: Another of the many possibilities, is to use a "First in, first out" sequence. (*In this sentence, the comma separates the subject, "Another," from the verb, "is."*)
Correct: Another of the many possibilities is to use a "first in, first out" sequence.

Incorrect: The schedules that have to be updated every month are, 14, 16, 21, 22, 27, and 31. (*In this sentence, the comma separates the verb from its complement.*)
Correct: The schedules that have to be updated every month are 14, 16, 21, 22, 27, and 31.

Incorrect: The company has grown so big, that an informal evaluation procedure is no longer effective. (*In this sentence, the comma separates the predicate adjective "big" from the clause that modifies it.*)

Correct: The company has grown so big that an informal evaluation procedure is no longer effective.

Incorrect: Recent studies, and reports by other firms confirm our experience. (*In this sentence, the comma separates the two elements in the compound subject.*)

Correct: Recent studies and reports by other firms confirm our experience.

Incorrect: New and old employees who use the processed order form, do not completely understand the basis of the system. (*In this sentence, a comma separates the subject and its restrictive modifier from the verb.*)

Correct: New and old employees who use the processed order form do not completely understand the basis of the system.

The Semicolon

Semicolons are used in the following instances.

- to separate independent clauses not linked by a coordinating conjunction

The second edition of the handbook is more up-to-date; however, it is more expensive.

- to separate items in a series that already contains commas

The members elected three officers: Jack Resnick, president; Carol Wayshum, vice president; Ahmed Jamoogian, recording secretary.

In this example, the semicolon acts as a "supercomma," keeping the names and titles clear.

Common error Using a semicolon when a colon is called for is not correct.

Incorrect: We still need to see one item; the RFP.
Correct: We still need to see one item: the RFP.

The Colon

Colons are used in the following instances.

- to introduce a word, phrase, or clause that amplifies, illustrates, or explains a general statement

The project team lacked one crucial member: a project leader.
Here is the client's request: we are to provide the preliminary proposal by November 13.
We found three substances in excessive quantities: potassium, cyanide, and asbestos.
The week had been productive: fourteen projects had been completed and another dozen had been initiated.

Note that the text preceding a colon should be able to stand on its own as a main clause:

Incorrect: We found: potassium, cyanide, and asbestos.
Correct: We found potassium, cyanide, and asbestos.

- to introduce items in a vertical list, if the sense of the introductory text would be incomplete without the list

We found the following:
potassium
cyanide
asbestos

- to introduce long or formal quotations

The president began: "In the last year . . ."

Common error Using a colon to separate a verb from its complement is not correct.

Incorrect: The tools we need are: a plane, a level, and a T-square.
Correct: The tools we need are a plane, a level, and a T-square.
Correct: We need three tools: a plane, a level, and a T-square.

The Dash

Dashes are used in the following instances.

- to set off a sudden change in thought or tone

The committee found—can you believe this?—that the company bore *full* responsibility for the accident.
That's what she said—if I remember correctly.

- to emphasize a parenthetical element

The managers' reports—all ten of them—recommend production cutbacks for the coming year.
Arlene Kregman—the first woman elected to the board of directors—is the next scheduled speaker.

- to set off an introductory series from its explanation

Wetsuits, weight belts, tanks—everything will have to be shipped in.

When a series *follows* the general statement, a colon replaces the dash:

Everything will have to be shipped in: wetsuits, weight belts, and tanks.

Note that typewriters and most computer keyboards do not have a key for the dash (although most word-processing programs have dashes as special characters). In typewritten or word-processed text, a dash is represented by two uninterrupted hyphens. No space precedes or follows the dash.

Common error Using a dash as an all-purpose substitute for other punctuation marks is not correct.

Incorrect: The regulations—which were issued yesterday—had been anticipated for months.
Correct: The regulations, which were issued yesterday, had been anticipated for months.

Incorrect: Many candidates applied—however, only one was chosen.
Correct: Many candidates applied; however, only one was chosen.

Parentheses

Parentheses are used in the following instances.

- to set off incidental information

Please call me (x3104) when you get the information.
Galileo (1546–1642) is often considered the father of modern astronomy.
H. W. Fowler's *Modern English Usage* (New York: Oxford University Press, 2nd ed., 1965) is still the final arbiter.

- to enclose numbers and letters that label items listed in a sentence

To transfer a call within the office, (1) place the party on HOLD, (2) press TRANSFER, (3) press the extension number, and (4) hang up.

Use both a left and a right parenthesis—not just a right parenthesis—in this situation.

Common error Using parentheses instead of brackets to enclose the writer's interruption of a quotation (see the discussion of brackets, page 232) is not correct.

Incorrect: He said, "The new manager (Farnham) is due in next week."
Correct: He said, "The new manager [Farnham] is due in next week."

The Hyphen

Hyphens are used in the following instances.

- in general, to form compound adjectives that precede nouns

general-purpose register
meat-eating dinosaur
chain-driven saw

Note that hyphens are not used after words that end in -ly:

newly acquired terminal

Also note that hyphens are not used when the compound adjective follows the noun:

The Woodchuck saw is chain driven.

Many organizations have their own preferences about hyphenating compound adjectives. Check to see if your organization has a preference.

- to form some compound nouns

editor-in-chief
president-elect

- to form fractions and compound numbers

one-half
fifty-six

- to attach some prefixes and suffixes

post-1945
frost-free

- to divide a word at the end of a line

We will meet in the pavil-
ion in one hour.

Whenever possible, avoid such breaks; they slow down the reader. When you do use them, check the dictionary to make sure you have divided the word *between* syllables.

The Apostrophe

Apostrophes are used in the following instances.

- to indicate the possessive case

the manager's goals
the supervisor's lounge
the employees' credit union
Charles's T-square

For joint possession, add the apostrophe and the *s* to only the last noun or proper noun:

Watson and Crick's discovery

For separate possession, add an apostrophe and an *s* to each of the nouns or pronouns:

Newton's and Galileo's ideas

Make sure you do not add an apostrophe or an *s* to possessive pronouns: *his, hers, its, ours, yours, theirs.*

- to form contractions

I've
can't
shouldn't
it's

The apostrophe usually indicates an omitted letter or letters. For example, *can't* is *can(no)t, it's* is *it (i)s.*

Some organizations discourage the use of contractions; others have no preference. Find out the policy your organization follows.

Common error Using the contraction *it's* in place of the possessive pronoun *its* is incorrect.

Incorrect: The company does not feel that the problem is it's responsibility.
Correct: The company does not feel that the problem is its responsibility.

Quotation Marks

Quotation marks are used in the following instances.

- to indicate titles of short works, such as articles, essays, or chapters

Smith's essay "Solar Heating Alternatives"

- to call attention to a word or phrase that is being used in an unusual way or in an unusual context

A proposal is "wired" if the sponsoring agency has already decided who will be granted the contract.

Don't use quotation marks as a means of excusing a poor word choice

The new director has been a real "pain."

- to indicate direct quotation—that is, the words a person has said or written

"In the future," he said, "check with me before authorizing any large purchases."
As Breyer wrote, "Morale *is* productivity."

Do not use quotation marks to indicate indirect quotation:

Incorrect: He said that "third-quarter profits would be up."
Correct: He said that third-quarter profits would be up.
Correct: He said, "Third-quarter profits will be up."

Related punctuation Note that if the sentence contains a *tag*—a phrase identifying the speaker or writer—a comma is used to separate it from the quotation:

John replied, "I'll try to fly out there tomorrow."
"I'll try to fly out there tomorrow," John replied.

Informal and brief quotations require no punctuation before the quotation marks:

She said "Why?"

In the United States (but not in most other English-speaking nations), commas and periods at the end of quotations are placed within the quotation marks:

The project engineer reported, "A new factor has been added."
"A new factor has been added," the project engineer reported.

Question marks, dashes, and exclamation points, on the other hand, are placed inside the quotation marks when they apply only to the quotation and outside the quotation marks when they apply to the whole sentence:

He asked, "Did the shipment come in yet?"
Did he say, "This is the limit"?

Note that only one punctuation mark is used at the end of a set of quotation marks:

Incorrect: Did she say, "What time is it?"?
Correct: Did she say, "What time is it?"

Block quotations When quotations reach a certain length—generally, more than four lines—writers tend to switch to a block format. In a typewritten manuscript, a block quotation is usually

- indented 10 spaces from the left-hand margin
- single-spaced
- typed without quotation marks
- introduced by a complete sentence followed by a colon

(Different organizations observe their own variations on these basic rules.)

McFarland writes:

> The extent to which organisms adapt to their environment is still being charted. Many animals, we have recently learned, respond to a dry winter with an automatic birth-control chemical that limits the number of young to be born that spring. This prevents mass starvation among the species in that locale.

Hollins concurs. She writes, "Biological adaptation will be a major research area during the next decade."

Mechanics

Ellipses

Ellipses (three spaced periods) indicate the omission of some material from a quotation. A fourth period with no space before it precedes ellipses when the sentence in the source has ended and you are omitting material that follows, or when the omission follows a portion of the source's sentence that is in itself a grammatically complete sentence:

> "Send the updated report . . . as soon as you can."
>
> Larkin refers to the project as "an attempt . . . to clarify the issue of compulsory arbitration . . . We do not foresee an end to the legal wrangling . . . but perhaps the report can serve as a definition of the areas of contention."

The second example has omitted words after "attempt" and after "wrangling." In addition, it has used a sentence period plus three spaced periods after "arbitration," which ends the original writer's sentence; and it has omitted the following sentence.

Brackets

Brackets are used in the following instances.

- to indicate words added to a quotation

"He [Pearson] argued against the proposal."

A better approach would be to shorten the quotation:

The minutes of the meeting note that Pearson "argued against the proposal."

- to indicate parentheses within parentheses

(For further information, see Charles Houghton's *Civil Engineering Today* [New York: Arch Press, 1993].)

Italics

If your typewriter or word processor does not have italic type, indicate italics by underlining.

Darwin's Origin of Species is still read today.

Italics are used in the following instances.

- for words used as words

In this report, the word *operator* will refer to any individual who is actually in charge of the equipment, regardless of that individual's certification.

- to indicate titles of long works (books, manuals, etc.), periodicals and newspapers, long films, long plays, and long musical works

See Houghton's *Civil Engineering Today*.
We subscribe to the *Commerce Business Daily*.

- to indicate the names of ships, trains, and airplanes

The shipment is expected to arrive next week on the *Penguin.*

- to set off foreign expressions that have not become fully assimilated into English

The speaker was guilty of *ad hominem* arguments.

- to emphasize words or phrases

Do not press the ERASE key.

Numbers

The use of numbers varies considerably. Therefore, you should find out what guidelines your organization or research area follows in choosing between words and numerals. Many organizations use the following guidelines.

- Use numerals for technical quantities, especially if a unit of measurement is included:

3 feet
12 grams
43,219 square miles
36 hectares

- Use numerals for nontechnical quantities of 10 or more:

300 persons
12 whales
35% increase

- Use words for nontechnical quantities of fewer than 10:

three persons
six whales

- Use both words and numerals:
  - for back-to-back numbers

six 3-inch screws
fourteen 12-foot ladders
3,012 five-piece starter units

In general, use the numeral for the technical unit. If the nontechnical quantity would be cumbersome in words, use the numeral.

  - or round numbers over 999,999

14 million light-years
$64 billion

  - for numbers in legal contracts or in documents intended for international readers

thirty-seven thousand dollars ($37,000)
five (5) relays

  - for addresses

3801 Fifteenth Street

Special Cases

- If a number begins a sentence, use words, not numerals:

Thirty-seven acres was the agreed-upon size of the lot.

Many writers would revise the sentence to avoid this problem:

The agreed-upon size of the lot was 37 acres.

- Don't use both numerals and words in the same sentence to refer to the same unit:

On Tuesday the attendance was 13; on Wednesday, 8.

- Write out fractions, except if they are linked to technical units:

two-thirds of the members
3½ hp

- Write out approximations:

approximately ten thousand people
about two million trees

- Use numerals for titles of figures and tables and for page numbers:

Figure 1
Table 13
page 261

- Use numerals for decimals:

  3.14
  1,013.065

  Add a zero before decimals of less than 1:

  0.146
  0.006

- Avoid expressing months as numbers, as in "3/7/94": in the United States, this means March 7, 1994; in most other countries, it means July 3, 1994. Use one of the following forms:

  March 7, 1994
  7 March 1994

- Use numerals for times if A.M. or P.M. is used:

  6:10 A.M.
  six o'clock

Abbreviations

Abbreviations provide a useful way to save time and space, but you must use them carefully; you can never be sure that your readers will understand them. Many companies and professional organizations have lists of approved abbreviations.

Analyze your audience in determining whether and how to abbreviate. If your readers include nontechnical people unfamiliar with your field, either write out the technical terms or attach a list of abbreviations. If you are new in an organization or are writing for publication for the first time in a certain field, find out what abbreviations are commonly used. If for any reason you are unsure about whether or how to abbreviate, write out the word.

The following are general guidelines about abbreviations.

- You may make up your own abbreviations. For the first reference to the term, write it out and include, parenthetically, the abbreviation. In subsequent references, use the abbreviation. For long works, you might want to write out the term at the start of major units, such as chapters.

  The heart of the new system is the self-loading cartridge (slc).

  This technique is also useful, of course, in referring to existing abbreviations that your readers might not know:

  The cathode-ray tube (CRT) is your control center.

- Most abbreviations do not take plurals:

  1 lb
  3 lb

- Most abbreviations in scientific writing are not followed by periods:

  lb
  cos
  dc

  If the abbreviation can be confused with another word, however, use a period:

  in.
  Fig.

- Spell out the unit if the number preceding it is spelled out or if no number precedes it:

How many square meters is the site?

Capitalization

For the most part, the conventions of capitalization in general writing apply in workplace writing.

- Capitalize proper nouns, titles, trade names, places, languages, religions, and organizations:

William Rusham
Director of Personnel
Quick Fix Erasers
Bethesda, Maryland
Methodism
Italian
Society for Technical Communication

In some organizations, job titles are not capitalized unless they refer to specific persons:

Alfred Loggins, Director of Personnel, is interested in being considered for vice president of marketing.

- Capitalize headings and labels:

A Proposal to Implement the Wilkins Conversion System
Section One
The Problem
Figure 6
Mitosis
Table 3
Rate of Inflation, 1980–1990

Appendix C

Commonly Misused Words and Phrases

This appendix explains the proper usage of some of the most commonly misused words and phrases in workplace writing. A brief list like this one cannot replace a full-length work: I recommend Harry Shaw's *Dictionary of Problem Words and Expressions* (McGraw-Hill, 1987) and Theodore M. Bernstein's *The Careful Writer: A Modern Guide to English Usage* (Atheneum, 1965).

In describing the proper usage of these words and phrases, I offer brief explanations and/or examples. The examples are enclosed within quotation marks.

accept, except. "We will not accept delivery of any items except those we have ordered."

amount, number. *Amount* is used for noncounting items; *number* refers to counting items: "the amount of concrete," but "the number of bags of concrete."

adapt, adopt. *Adapt* means to adjust or to modify; *adopt* means to accept. "Management decided to adapt the quality-circle plan rather than adopt it as is."

affect, effect. *Affect* is a verb: "How will the news affect him?" *Effect* is most commonly a noun: "What will be the effect of the increase in allowable limits?" *Effect* is also (rarely) a verb meaning to bring about or cause to happen: "The new plant is expected to effect a change in our marketing strategy."

already, all ready. "The report had already been sent to the printer when the writer discovered that it was not all ready."

alright, all right. *Alright* is a misspelling of *all right.*

among, between. In general, *among* is used for relationships of more than two items; *between* is used for only two items. "The collaboration among the writer, the illustrator, and the printer," but "the agreement between the two companies."

assure, ensure, insure. *To assure* means to put someone's mind at ease: "let me assure you." *To ensure* and *to insure* both mean to make sure: "the new plan will ensure [or insure] good results." Some writers prefer to use *insure* only when referring to insurance: "to insure against fire loss."

can, may, might. *Can* refers to ability: "We can produce 300 chips per hour." *May* refers to permission: "May I telephone your references?" *Might* refers to possibility: "We might see further declines in PC prices this year."

could of. This is not a correct phrase; it is a corruption of *could've,* the contraction of *could have:* "She could have mentioned the abrasion problem in the progress report."

compliment, complement. A *compliment* is a statement of praise: "The owner offered a gracious compliment to the architect on his design." The word is also a verb: "The owner graciously complimented the architect." A *complement* is something that fills something up or makes it complete, or something that is an appropriate counterpart: "The design is a perfect complement to the landscape."

The word is also a verb: "The design complements the landscape perfectly."

criteria, criterion. *Criteria,* meaning standards against which something will be measured, is plural; *criterion* is singular.

data, datum. *Data* is plural; *datum* is singular. However, the distinction is fading in popular usage, although not in some scientific and engineering applications. My advice: check to see how it is used in your company or field.

discreet, discrete. *Discreet* means careful and prudent: "She is a very discreet manager; you can confide in her." *Discrete* means separate or distinct: "The company will soon split into three discrete divisions."

effective, efficient. *Effective* means that the item does what is is meant to do; *efficient* also carries the sense of accomplishing the goal without using more resources than necessary. "Air Force 1 is an effective way to move the president around, but it is not efficient; it costs some $40,000 per hour to fly."

either . . . or; neither . . . nor. *Either . . . or* means one of two; *neither . . . nor* means not one of two: "Either Jim or I will attend the meeting, but neither Bob nor Ahmed will."

farther, further. *Farther* refers to distance: "one mile farther down the road." *Further* means greater in quantity, time, or extent: "Are there any further questions?"

feedback. Many writers will not use *feedback* to refer to a response by a person: "Let me have your feedback by Friday." They limit the term to its original meaning, dealing with electricity, because a human response involves thinking (or should, anyway).

fewer, less. *Fewer* is used for counting items: "fewer bags of cement;" *less* is used for noncounting items: "less cement." It's the same distinction as between *number* and *amount.*

foreword, forward. A *foreword* is a preface, usually written by someone other than the author, introducing and praising the book. *Forward* refers to being in advance: "The company decided to move forward with the project."

i.e., e.g. *I.e.,* Latin for *id est,* means *that is. E.g.,* Latin for *exempli gratia,* means *for example.* Writers who don't know Latin (like myself) often confuse them. That's why I recommend using the English versions. Also, add commas after them: "Use the English versions—that is, *that is* and *for example.*"

imply, infer. The writer or speaker implies; the reader or listener infers.

input. People who don't like to give their *feedback* also don't like to offer their *input.*

its, it's. *Its* is the possessive pronoun: "The lab rat can't make up its mind." *It's* is the contraction of *it is:* "It's too late to apply for this year's grant." People mix up these two words because they remember learning that possessives take apostrophes—"Bob's computer"—so when they use the possessive form of it, they add the apostrophe. But *its* is a possessive pronoun, like *his, hers, theirs, ours* and *yours,* a word specifically created to fulfill only one function: to indicate possession. It is not the possessive form of another word, and therefore it does not take an apostrophe.

-ize. There are many legitimate words ending in *ize,* such as *harmonize* and *sterilize,* but many writers and readers can't stand new ones (such as *prioritize*) when there are pefectly fine words already (such as *rank*).

lay, lie. *Lay* is a transitive verb meaning to place: "Lay the equipment on the table." *Lie* is an intransitive verb meaning to recline: "Lie down on the couch." The complete conjugation of *lay* is *lay, laid, laid, laying;* of *lie,* it is *lie, lay, lain, lying.*

lead, led. *Lead* is the infinitive verb: "We want to lead the industry." *Led* is the past tense: "Last year we led the industry."

parameter. This is a mathematical term referring to a constant whose value can vary according to its application. Many writers object to the nonmathematical uses of the term, including such concepts as perimeter, scope, outline, and limit. (You guessed it: the same people who don't provide input or feedback don't use parameter very much either.)

phenomena, phenomenon. *Phenomena* is plural; *phenomenon* is singular.

plain, plane. *Plain* means simple and unadorned: "The new company created a very plain logo." *Plane* has several meanings: an airplane, the act of smoothing a surface, the tool used to smooth a surface, and the flat surface itself.

precede, proceed. *Precede* means to come before: "Should Figure 1 precede Figure 2?" *Proceed* means to move forward: "We decided to proceed with the project despite the setback."

shall, will. *Shall* is used to suggest a legal obligation, particularly in a formal specification or contract: "The contractor shall remove all existing rebar." *Will* does not suggest a legal obligation: "We will get in touch with you as soon as possible to schedule the job."

sight, site, cite. *Sight* refers to vision; *site* is a place; *cite* is a verb meaning to document a reference.

than, then. *Than* is a conjunction used in comparisons: "Plan A works better than Plan B." *Then* is an adverb referring to time: "First we went to the warehouse. Then we went to the plant."

their, there, they're. *Their* is the possessive pronoun: "They brought their equipment with them." *There* is used to refer to a place—"We went there yesterday"—or in expletive expressions—"There are three problems we have to solve." *They're* is the contraction of *they are.*

to, too, two. *To* is used in infinitive verbs ("to buy a new microscope") and in expressions referring to direction ("go to Detroit"). *Too* means excessively: "The refrigerator is too big for the lab." *Two* is the number 2.

viable. This is a fine Latin word meaning able to sustain life: "viable cell culture" and "viable fetus." Many writers avoid such clichés as *viable alternative* (while they're avoiding *input* and *feedback*).

weather, whether. *Weather* refers to sunshine and temperature. *Whether* refers to alternatives. "The demonstration will be held outdoors whether or not the weather cooperates."

who's, whose. *Who's* is the contraction of *who is. Whose* is the possessive case of *who*: "Whose printer are we using?"

-wise. As in *job retentionwise, accelerationwise,* or *RAMwise.* See the entry on *-ize.*

Xerox. The people at Xerox become unhappy when writers ask for a xerox copy (unless it's made on a Xerox copier). The correct word is *photocopy; Xerox* is a copyrighted term.

your, you're. *Your* is the possessive pronoun: "Bring your calculator to the meeting." *You're* is the contraction of *you are.*

Appendix D

Guidelines for Speakers of English as a Second Language

English is notoriously difficult to master, and no brief guide can answer all your questions about the many eccentricities of the language. The purpose of this appendix is to point out ten aspects of writing in English that are most difficult for nonnative speakers.

The advice offered here is based on a highly regarded book, Ann Raimes' *Grammar Troublespots: An Editing Guide for ESL Students* (St. Martin's Press, 1988). It is an excellent investment.

Basic Characteristics of a Sentence

A sentence has five characteristics:

- It starts with a capital letter and ends with a period, a question mark, or (rarely) an exclamation point.

- It has a subject, usually a noun. The subject is what the sentence is about.
- It has a verb, which tells the reader about what happens to the subject.
- It has a standard word order. The most common sequence in English is subject-verb-object. But you can add information at different locations. For instance, here is a sentence in the basic subject-verb-object order:

  We hired a consulting firm.

  You can add information at the start of the sentence:

  Yesterday we hired a consulting firm.

  Or you can add information at the end of the sentence:

  Yesterday we hired a consulting firm, Sanderson & Associates.

  Or, you can add information in the middle:

  Yesterday we signed a nontransferrable contract with a consulting firm, Sanderson & Associates.

  In fact, any element of a sentence can be expanded.
- It has an independent clause, an idea that can stand alone. The following is a sentence because its idea can stand alone:

  The pump failed because of improper maintenance.

  The following is also a sentence:

The pump failed.

But the following is *not* a sentence:

Because of improper maintenance.

Linking Ideas by Coordination

One way to connect ideas is by coordination. Coordination means the two ideas are roughly equal in importance. There are four main ways to coordinate ideas:

- Use a semicolon to coordinate two sentences with similar structures:

  The information for bid was published last week; the proposal is due in less than a month.

- Use a comma and one of the coordinating conjuctions (*and, but, or, nor, so, for,* and *yet*) to coordinate two ideas.

  The information for bid was published last week, but the proposal is due in less than a month.

  In this example, the *but* clarifies the relationship between the two clauses: you haven't been given as much time as you need to write the proposal.
- Combine two separate sentences into one. Here are two separate sentences:

  The bridge was completed last year. The bridge already needs repairs.

  One way to combine them would be as follows:

The bridge was completed last year and already needs repairs.

Notice that there is no comma after *year* because the two verbs in the sentence have the same subject.

- Use transitional words and phrases within and between sentences (see Chapter 6 for more information on transitional words and phrases).

The 486 chip has already replaced the 386. *In fact,* it's hard to find a 386 in a new computer.

Linking Ideas by Subordination

Two separate sentences can also be linked by subordination, that is, by de-emphasizing one of them. There are two basic ways to use subordination:

- *Combine two separate sentences into one.* Here are two separate sentences from the preceding section:

The bridge was completed last year. The bridge already needs repairs.

Here are the two combined:

Completed last year, the bridge already needs repairs.

In this version, *completed last year* modifies *the bridge.* The independent clause is *the bridge already needs repairs.*

- *Use a subordinating word or phrase.* Start with the two separate sentences about the bridge; then combine them as follows:

The bridge, which was completed last year, already needs repairs.

This version emphasizes the *already needs repairs* portion of the sentence and deemphasizes the *was completed last year* portion by putting it in a *which* clause. Another way to subordinate is as follows:

The bridge already needs repairs, although it was completed last year.

In this example, *although* subordinates the clause, leaving *the bridge already needs repairs* as the independent clause. Note that the order of the sentences could be reversed:

Although it was completed last year, the bridge already needs repairs.

In general, it is easier to read a sentence if the subject (*the bridge*) appears before the pronoun that replaces it (*it*). This way, readers don't have to remember the *it* clause until they find out what the subject of the sentence is.

Verb Tenses

Verb tenses in English can be complicated, but in general there are four kinds of time relationships that you need to understand.

- simple past, present, and future

  Examples:

  Yesterday we *subscribed* to a new IEEE journal.
  We *subscribe* to three IEEE journals. (Meaning: we subscribe to the three journals regularly.)
  We *will subscribe* to the new IEEE journal. (Or: We *are going* to subscribe to the new IEEE journal.)

- an action in progress at a known time

  Examples:

  We *were updating* our directory when the power failure occurred. (Meaning: after the power was restored, we continued.)

We *are updating* our directory now.
We *will be updating* our directory tomorrow when you arrive.

- an action completed before a known time

Examples:

We *had started* to write the proposal when we got your call.
We *have started* to write the proposal.
We *will have started* to write the proposal by the time you arrive. (Meaning: both events are in the future, with the writing beginning before the arrival.)

- an action in progress until a known time

Examples:

We *had been working* on the reorganization when the news of the merger was publicized. (Meaning: after the news, we stopped working on it.)
We *have been working* on the reorganization for over a year. (Meaning: the work will continue into the future.)
We *will have been working* on the reorganization for two years by the time the reorganization occurs.

Agreement Between the Subject and the Verb

The subject and the verb in a clause or sentence must agree in number. There are five major constructions to remember:

- simple agreement

Examples:

The new valve is installed according to the manufacturer's specifications.
The new valves are installed according to the manufacturer's specifications.

- agreement when the clause or sentence contains information between the subject and the verb

  Examples:

  The result of the tests is included in Appendix C.
  The results of the test are included in Appendix C.

- Agreement when the clause or sentence contains special pronouns and quantifiers. Pronouns that end in *-body* or *-one*—such as *everyone, everybody, someone, somebody, anyone, anybody, no one,* and *nobody*—are singular. In addition, quantifiers such as *something, each,* and *every* are singular.

  Examples:

  Everybody is invited to the pre-proposal meeting.
  Each of the members is asked to submit billable hours by the end of the month.

- agreement when the clause or sentence contains a compound subject. In such cases, the verb must be plural.

  Example:

  The contractor and the subcontractor want to meet to resolve the difficulties.

- agreement when a relative pronoun such as *who, that,* or *which* begins a clause. In such cases, make sure the verb agrees in number with the noun that the relative pronoun refers to.

  Examples:

  The *numbers* that *are* used in the formula *do* not agree with *the ones* we were given at the site.
  The *number* that *is* used in the formula *does* not agree with *the one* we were given at the site.

Articles

Few aspects of English can be as frustrating to the nonnative speaker of English as the simple words *a, an*, and *the*. Although there are a few rules that you should try to learn, remember that there are many exceptions and special cases. Here are three general guidelines.

- Singular proper nouns—those that name specific persons, places, and things—do not usually take an article.

  Examples:

  Taiwan
  James Allenby

  But plural proper nouns often do take an article:

  Examples:

  the United States
  the Allenbys

- Countable common nouns take an article:

  Examples:

  the microscope
  a desk

  But uncountable common nouns generally do not take an article:

  Examples:

  overtime
  equipment
  integrity

  How can you be sure if a word is countable or uncountable? Unfortunately, you can't. You have to keep a list.

- Common nouns can be referred to as either specific or nonspecific. The specific form takes *the*; the nonspecific form takes either *a* or *an*. Here is a sentence in which the reference is specific:

  Example:

  We need to get back to *the experiment.*

  The writer assumes that the reader understands which experiment is being referred to.

  Here is a sentence in which the reference is nonspecific:

  Example:

  We need to begin *an experiment.*

Adjectives

There are three main points to keep in mind about adjectives in English:

- Adjectives do not take a plural form.

  Examples:

  a complex project
  two complex projects

- Adjectives can be placed before the nouns they modify or later in the sentence.

  Examples:

  The critical need is to reduce the drag coefficient.
  The need to reduce the drag coefficient is critical.

- Adjectives of one or two syllables take special endings to create the comparative and superlative forms.

  Examples:

  big, bigger, biggest
  heavy, heavier, heaviest

  Adjectives of three or more syllables take the word *more* for the comparative form and the words *the most* for the superlative form.

  Examples:

  qualified, more qualified, the most qualified
  feasible, more feasible, the most feasible

Adverbs

Adverbs, like adjectives, are modifiers, but they are somewhat more complex in their placement in the sentence. Remember four points about adverbs:

- Adverbs modify verbs.

  Example:

  Management terminated the project *reluctantly.*

- Adverbs also modify adjectives.

  Example:

  The executive summary was *conspicuously* absent.

- Adverbs that describe how an action takes place can be placed in different locations in the sentence.

Examples:

Carefully the inspector examined the welds.
The inspector *carefully* examined the welds.
The inspector examined the welds *carefully*.

But don't place the adverb between the verb and the direct object.

Example:

The inspector examined *carefully* the welds.

- Adverbs that describe the whole sentence can also be placed in different locations in the sentence.

  Examples:

  Apparently, the inspection was successful.
  The inspection *apparently* was successful.
  The inspection was successful, *apparently*.

–ing Forms of Verbs

English uses the –ing form of verbs in four major ways:

- as part of a verb in a sentence

  Example:

  We are *shipping* the materials by UPS.

- to add extra information in a sentence

  Examples:

  Analyzing the sample, we discovered two anomalies.

The sample *containing* the anomolies appears on slide 14.
We studied the sample, *thinking* it could be important.

- to serve as an adjective

  Example:

  the *leaking* pipe

- to serve as a noun

  Example:

  Writing is the best way to learn to write.

Conditions

There are four main types of condition used with the word *if* in English:

- conditions of fact

  Examples:

  If you see "Unrecoverable Application Error," the program has crashed.
  If rats are allowed to eat as much as they want, they become obese.

- future prediction

  Example:

  If we win this contract, we will have to add three more engineers.

- present-future speculation

Example:

If I were president of the company, I would be much more aggressive.

Notice that the present-future speculation usage suggests a condition contrary to fact. For instance, in this example, the implication is that you are not president of the company. Because the sentence describes a condition contrary to fact, the writer uses "were" rather than "was" in the first clause.

- past speculation

  Example:

  If we had won this contract, we would have had to add three more engineers.

- Notice, here, too that the implication is that the condition is contrary to fact.

One other point about *if* conditions: you can restructure the sentence and not use *if* at all:

Example:

Had we won this contract, we would have had to add three more engineers.

Appendix E

Guidelines for Writing to Speakers of English as a Second Language

Many people know that English is the international langauge, but few know that of the 600 million people who speak English, more than half did not learn it as their first language (Martin and Chaney 1992). More and more often we find ourselves writing to people who are not fully fluent in English. The following guidelines, based on a discussion by Harris and Moran (1987), can help you avoid some common pitfalls.

1. Try to use the most common words of English. Use "improve" rather than "meliorate."
2. Do not create a new word by using a common word in an uncommon way.

Weak: Let me caveat the advice I gave you in my last letter.
Better: Let me add a cautionary statement about the advice I gave in my last letter.

3. Use clear, action-specific phrases. Instead of "making contact with" someone, "telephone" the person.
4. Avoid complicated phrases.

Weak: New hardware has reached the marketplace to replace microprocessors in neural network applications.
Better: New hardware is now available to replace microprocessors in neural network applications.

5. Learn basic spelling differences between American English and British English, which is used in most other countries. Examples: *color* versus *colour* and *organization* versus *organisation.*
6. Use standard grammar, and do not leave out information that might be implicit for a native speaker. The nonnative speaker of English relies on all the words to understand the idea.

Weak: Send the original calculations. Not the updates.
Better: Send the original calculations; do not send the updates.

7. Do not leave out punctuation.

Weak: If the process is done properly the result is a homogeneous metal structure with greater strength.
Better: If the process is done properly, the result is a homogeneous metal structure with greater strength.

8. Avoid figures of speech, especially clichés.

Weak: We want to hit the ground running when the product is released in January.
Better: We want to be fully prepared when the product is released in January.

9. Avoid sports expressions, for most nonnative readers will not be familair with them.

Weak: The best marketing strategy for the new chip is a full court press.
Better: The best marketing strategy for the new chip is to advertise extensively.

10. Avoid an overly familiar tone. Use last names rather than first names unless you know the person very well.

Weak: The best way to go, Chaman, is to try nondestructive testing first.
Better: We recommend that you try nondestructive testing first, Mr. Sahni.

As you have probably noticed, these simple guidelines are very similar to basic principles of good writing to *any* reader. The most challenging aspect of writing to speakers of English as a second language is to understand the differences between your culture and theirs. This kind of information varies from country to country, of course, but in general it concerns such issues as the role of negotiations, the nature of rules and regulations, the role of technology, and even the nature of time. I recommend an article by Limaye and Victor (1991) for a concise overview of these issues. Also, see Harris and Moran (1987) for a full-length treatment of the implications of cultural differences for managers.

References

Harris, P. R., and R. T. Moran. 1987. *Managing cultural differences.* 2nd ed. Houston: Gulf.

Limaye, M. R., and D. A. Victor. 1991. Cross-cultural business communication research: State of the art and hypotheses for the 1990s. *Journal of Business Communication* 28, no. 3: 277–99.

Martin, J. S., and L. H. Chaney. 1992. Determination of content for a collegiate course in intercultural business communication by three Delphi panels. *Journal of Business Communication* 29, no. 3: 267–83.

Appendix F

Selected Bibliography

Ethics

Beauchamp, T. L., and N. E. Bowie. *Ethical Theory and Business.* 4th ed. Englewood Cliffs, N.J.: Prentice Hall, 1993.

Behrman, J. N. *Essays on Ethics in Business and the Professions.* Englewood Cliffs, N.J.: Prentice Hall, 1988.

Brockmann, R. J. and F. Rook, eds. *Technical Communication and Ethics.* Washington, D.C.: Society for Technical Communication, 1989.

Chalk, R. *AAAS Professional Ethics Project: Professional Ethics Activities in the Scientific and Engineering Societies.* Washington, D.C.: American Association for the Advancement of Science, 1980.

Honor in Science. New Haven, Conn.: Sigma Xi. The Scientific Research Society, 1986.

Velasquez, M. G. *Business Ethics: Concepts and Cases.* 3rd ed. Englewood Cliffs, N.J.: Prentice Hall, 1992.

Workplace Writing

Beer, D. F. *Writing and Speaking in the Technology Professions: A Practical Guide.* New York: IEEE, 1992.

Bell, P. *High Tech Writing: How to Write for the Electronics Industry.* New York: Wiley-Interscience, 1986.

Blicq, R. S. *Technically—Write! Communicating in a Technological Era.* 4th ed. Englewood Cliffs, N.J.: Prentice Hall, 1993.

Brusaw, C. T., G. J. Alred, and W. E. Oliu. *Handbook of Technical Writing.* 4th ed. New York: St. Martin's, 1991.

Kolin, P., and J. Kolin. *Models for Technical Writing.* New York: St. Martin's, 1985.

Markel, M. H. *Technical Writing: Situations and Strategies.* 3rd ed. New York: St. Martin's, 1992.

Mathes, J. C., and D. W. Stevenson. *Designing Technical Reports: Writing for Audiences in Organizations.* 2nd ed. Indianapolis: Bobbs-Merrill, 1991.

Pickett, N. A., and A. A. Laster. *Technical English.* 6th ed. New York: HarperCollins, 1993.

Price, J., and H. Korman. *How to Communicate Technical Information: A Handbook of Software and Hardware Documentation.* Redwood City, Calif.: Benjamin/Cummings, 1993.

Sherman, T. A., and S. S. Johnson. *Modern Technical Writing.* 5th ed. Englewood Cliffs, N.J.: Prentice Hall, 1990.

Also see the following journals:

IEEE Transactions on Professional Communication

Journal of Business and Technical Communication

Journal of Technical Writing and Communication

Technical Communication

Technical Communication Quarterly

Usage and General Writing

Bernstein, T. M. *The Careful Writer: A Modern Guide to English Usage.* New York: Atheneum, 1977.

Corbett, E. P. J. *Classical Rhetoric for the Modern Student.* 3rd ed. New York: Oxford University, 1990.

Flesch, R. *The Art of Plain Talk.* New York: Macmillan, 1988.

Fowler, H. W. *A Dictionary of Modern English Usage.* 2d ed., rev. by Sir E. Gowers. New York: Oxford University, 1987.

Maggio, R. *The Dictionary of Bias-Free Usage: A Guide to Nondiscriminatory Language.* Phoenix, Ariz.: Oryx, 1991.

Sorrels, B. D. *The Nonsexist Communicator: Solving the Problems of Gender and Awkwardness in Modern English.* Englewood Cliffs, N.J.: Prentice Hall, 1983.

Strunk, W., and E. B. White. *The Elements of Style.* 3rd ed. New York: Macmillan, 1979.

Williams, J. *Style: Ten Lessons in Clarity and Grace.* 3rd ed. Glenview, Ill.: Scott Foresman, 1989.

Graphics

Arntson, A. E. *Graphic Design Basics.* New York: Holt, Rinehart & Winston, 1988.

Baird, R. N., et al. *The Graphics of Communication.* 6th ed. San Diego: Harcourt Brace Jovanovich, 1992.

Crow, W. C. *Communication Graphics.* Englewood Cliffs, N.J.: Prentice Hall, 1986.

Earle, J. H. *Graphics for Engineers.* 2nd ed. Reading, Mass.: Addison-Wesley, 1988.

Foley, J. D. *Computer Graphics: Principles and Practice.* 2nd ed. Reading, Mass.: Addison-Wesley, 1990.

Hoffman, E. K., and J. Teeple. *Computer Graphics Applications: An Introduction to Desktop Publishing and Design, Presentation Graphics, Animation.* Belmont, Calif.: Wadsworth, 1990.

Lefferts, R. *How to Prepare Charts and Graphs for Effective Reports.* New York: Harper & Row, 1982.

Morris, G. E. *Technical Illustrating.* Englewood Cliffs, N.J.: Prentice Hall, 1975.

Smith, R. C. *Basic Graphic Design.* Englewood Cliffs, N.J.: Prentice Hall, 1986.

Talman, M. *Understanding Presentation Graphics.* San Francisco: Sybex, 1992.

Tufte, E. R. *Envisioning Information.* Cheshire, Conn.: Graphics Press, 1990.

Tufte, E. R. *The Visual Display of Quantitative Information.* Cheshire, Conn.: Graphics Press, 1983.

Wileman, R.E. *Visual Communicating.* Englewood Cliffs, N.J.: Education Technology Publications, 1993.

Also see the following journals:

Graphic Arts Monthly
Graphics: USA

Technical Manuals

Brockmann, R. J. *Writing Better Computer Documentation: From Paper to Hypertext.* New York: Wiley-Interscience, 1990.

Cohen, G., and D. H. Cunningham. *Creating Technical Manuals: A Step-by-Step Approach to Writing User-Friendly Instructions.* New York: McGraw-Hill, 1984.

Forbes, M. *Writing Technical Articles, Speeches, and Manuals.* 2nd ed. New York: Krieger, 1992.

Schoff, G. H., and Robinson, P. A. *Writing and Designing Manuals.* 2nd ed. Boca Raton, Fla.: Lewis, 1991.

Slatkin, E. *How to Write a Manual.* Berkeley, Calif.: Ten Speed, 1991.

Weiss, E. H. *How to Write Usable User Documentation.* 2nd ed. Phoenix, Ariz.: Oryx, 1992.

Engineering Specifications

Ayers, C. *Specifications for Architecture, Engineering, and Construction.* 2nd ed. New York: McGraw-Hill, 1984.

Goldblum, J. *Engineering Construction Specifications: The Road to Better Quality, Lower Cost, Reduced Litigation.* New York: Van Nostrand Reinhold, 1989.

Purdy, D. C. *A Guide to Writing Successful Engineering Specifications.* New York: McGraw-Hill, 1991.

Rosenfeld, W. *The Practical Specifier: A Manual of Construction Documentation for Architects.* New York: McGraw-Hill, 1985.

Proposals

Bowman, J. P., and B. P. Branchaw. *How to Write Proposals that Produce.* Phoenix, Ariz.: Oryx, 1992.

Hegelson, D. V. *Handbook for Writing Technical Proposals that Win Contracts.* Englewood Cliffs, N.J.: Prentice Hall, 1986.

Hill, J. W., and T. Whalen. *How to Create and Present Successful Government Proposals.* New York: IEEE, 1993.

Lefferts, R. *Getting a Grant in the 1990s: How to Write Successful Grant Proposals.* Englewood Cliffs, N.J.: Prentice Hall, 1991.

Meador, R. *Guidelines for Preparing Proposals.* 2nd ed. Boca Raton, Fla. Lewis, 1991.

Society for Technical Communication. *Proposals and their Preparation.* Vol. 1. Washington, D.C.: Society for Technical Communication, 1973.

Stewart, R. D., and A. L. Stewart. *Proposal Preparation.* 2nd ed. New York: Wiley-Interscience, 1992.

Whalen, T. *Writing and Managing Winning Technical Proposals.* Norwood, Mass.: Artech House, 1987.

Style Manuals

American National Standards, Inc. *American National Standard for the Preparation of Scientific Papers for Written or Oral Presentation.* ANSI Z39.16—1972. New York: American National Standards Institute, 1979.

CBE Style Manual Committee. *Council of Biology Editors Style Manual: A Guide for Authors, Editors, and Publishers in the Biological Sciences.* 5th ed. Washington, D.C.: Council of Biology Editors, 1983.

The Chicago Manual of Style. 14th ed., rev. Chicago: University of Chicago, 1993.

Dodd, J. S., ed. *The ACS Style Guide: A Manual for Authors and Editors.* Washington, D.C.: American Chemical Society, 1986.

Pollack, G. *Handbook for ASM Editors.* Washington, D.C.: American Society for Microbiology, 1977.

Publications Manual of the American Psychological Association. 3nd ed. Washington, D.C.: American Psychological Association, 1983.

Rubens, P., ed. *Science and Technical Writing: A Manual of Style.* New York: Henry Holt, 1992.

Skillin, M., and R. Gay. *Words into Type.* 3rd ed. Englewood Cliffs, N.J.: Prentice Hall, 1986.

U.S. Government Printing Office Style Manual. Rev. ed. New York: Outlet, 1988.

Also, many private corporations, such as John Deere, DuPont, Ford Motor Company, General Electric, and Westinghouse, have their own style manuals.

Word Processing and Desktop Publishing

Banks, M. A., and A. Dibell. *Word Processing Secrets for Writers.* Cincinnati, Ohio.: Writer's Digest, 1989.

Burke, C. *Type from the Desktop: Designing with Type and Your Computer.* Chapel Hill, N.C.: Ventana, 1990.

Chicago Guide to Preparing Electronic Manuscripts. Chicago: University of Chicago Press, 1987.

Gosney, M., et al. *The Gray Book: Designing in Black and White on Your Computer.* Chapel Hill, N.C.: Ventana, 1990.

Krull, R. *Word Processing for Technical Writers.* Amityville, N.Y.: Baywood, 1988.

Muehlman, S. *Word Processing on Microcomputers: Applications and Exercises.* Englewood Cliffs, N.J.: Prentice Hall, 1989.

Parker, R. C. *Looking Good in Print: A Guide to Basic Design for Desktop Publishing.* 2nd ed. Chapel Hill, N.C.: Ventana, 1990.

Shushan, R., et al. *Desktop Publishing by Design.* Redmond, Wash.: Microsoft, 1991.

Sudol, R. A. *Textfiles: A Rhetoric for Word Processing.* New York: Harcourt Brace Jovanovich, 1987.

Zinsser, W. *Writing with a Word Processor.* New York: Harper & Row, 1983.

Index

A

B

C

D

E

F

N

O

P

Q

R

S

T

U

V

W